Fourier Transforms in Radar and Signal Processing

For a listing of recent titles in the *Artech House Radar Library*,
turn to the back of this book.

Fourier Transforms in Radar and Signal Processing

David Brandwood

Artech House
Boston • London
www.artechhouse.com

Library of Congress Cataloging-in-Publication Data
Information for this book is on file with the Library of Congress.

British Library Cataloguing in Publication Data
Brandwood, David.
Fourier transforms in radar and signal processing. — (Artech House radar library)
1. Signal processing—Mathematics 2. Radar—Mathematics 3. Fourier transformations I. Title
621.3'822'01515723

ISBN 1-58053-174-1

Cover design by Yekaterina Ratner

685 Canton Street
Norwood, MA 02062

International Standard Book Number: 1-58053-174-1

10 9 8 7 6 5 4 3 2 1

Contents

Preface

The basic material for this book has been accumulated from time to time over the author's working lifetime of about 40 years. The rules-and-pairs approach to Fourier transforms has been employed with good effect in a wide variety of problems, from pulse Doppler radar spectra to delay compensation, and antenna array patterns to efficient clutter simulation. It has been found generally easy and effective, quickly yielding useful results and allowing the user to see clearly the relationships between functions and transforms, waveforms and spectra, rather than losing sight of these in the complexities of integration. It seemed that the benefits of this approach should be better known, and the initial intention was to produce a technical note for use by the author's colleagues and successors. However, the interest shown and encouragement given by Artech House have been gratefully received, and the opportunity to publicize the technique more widely has been taken.

The support of Roke Manor Research in providing the facilities and freedom to write this book is gratefully acknowledged, as are the backing of C. J. Tarran and the reviewing of S. H. W. Simpson. The final acknowledgment is to the publisher's reviewer, remaining anonymous, who provided encouragement and useful comments.

1

Introduction

1.1 Aim of the Work

The Fourier transform is a valuable theoretical technique, used widely in fields such as applied mathematics, statistics, physics, and engineering. However, the relationship between a function and its transform is given by an integral, and a certain amount of tedious integration may be required to obtain the transform in a given application. In general, the user of this mathematical tool is interested in the functions and their transforms and not in the process of obtaining one from the other, which, even if not difficult, can be complicated and require care to avoid any small slip that might lead to an error in the result. If the transform function could be obtained without integration, this would be welcomed by most users. In fact, anyone performing many transforms in a particular field, such as radar, where the spectra corresponding to various, perhaps rather similar, waveforms are required, would notice that certain waveforms have certain transforms and that certain relationships between waveforms lead to corresponding relationships between spectra. With the knowledge of a relatively small number of waveform-transform pairs and the rules for combining and scaling transforms, a very substantial amount of Fourier transform analysis can be carried out without any explicit integration at all—the integrations are prepackaged within the set of rules and pairs.

The aim of this book is to represent the rules-and-pairs approach to Fourier transforms, first defined systematically by Woodward [1], and to illustrate its use. The rules and pairs are given in Chapter 2, and the remaining chapters employ the technique in a number of areas of application. These are mainly in the field of signal or waveform processing (except in Chapter 7), but the technique is general, of course, and this choice should certainly not imply that other users of the Fourier transform should not also find the technique of interest and value.

The aim is not to provide a handbook of solutions to particular problems in the areas covered, though some results may be found particularly interesting and useful. More specifically, it is to show how such problems might be tackled and how the technique can be used with ingenuity in a variety of ways. These produce results that would be more difficult and tedious to obtain by integration, and which, using integration, may be expressed in a form that is more difficult to interpret and understand. In fact, an important advantage of the method, aided by the notation used, is the greater clarity regarding the nature of the transforms, obtained by keeping attention on the functions rather than on the mechanics of integration. While the illustrations may not include a reader's particular problem, some examples may well be close enough to suggest how the problem could be tackled by the rules-and-pairs method and perhaps solved more easily than otherwise. Once the user has become familiar with the method, many results can be obtained surprisingly easily and concisely, and it is surprising how complicated some of the problems are that the method can be applied to.

1.2 Origin of the Rules and Pairs Method for Fourier Transforms

With the arrival of the technology of electronics early in the last century, the possibilities for handling information—sending, receiving, and processing it—expanded immensely over that which was possible with the mechanical technology, however ingenious, of the nineteenth. The need to understand the limits of performance, whatever the technology available, and the dependence of performance on various parameters, such as bandwidth and signal-to-noise ratio, led to the rise of the subject of information theory. Under the stimulus of war, a new application of electronics, radar, developed rapidly in the 1930s and 1940s and again theoretical analysis followed. In 1953, P. M. Woodward's monograph *Probability and Information Theory, with Applications to Radar* [1] appeared. The topics of radar detection, accuracy, resolution, and ambiguity in the later chapters are generally the reasons for the references to this book in modern radar textbooks, but, leading up to his conclusions, Woodward needed results in the field of waveform analysis, and this is the subject of his Chapter 2, in which his rules and pairs are introduced. Waveforms and spectra are connected, of course, by a Fourier transform relationship, and this technique is the principal tool of the "time-and-frequency analysis" that Woodward implies is the basis of much of the mathematical study of information theory, radio, and radar.

It is not claimed that Woodward's rules and pairs themselves are particularly original. The pairs are well known and, as Woodward says, the rules are "known by heart by most circuit mathematicians" [1]. What is perhaps new is the careful specification of the set of rules and the committed and consistent use of them to obtain transforms very directly and concisely. In addition to the results required for the mathematical study of radar, Woodward uses this approach to derive important general results very neatly (again already known) such as Parseval's theorem, sampling theorems, and Poisson's formula. What is more clearly new and valuable is Woodward's contribution to notation, including the rect and sinc functions, the comb function and the rep operator. The term *sinc function* has since become more widely accepted and used, although, regrettably, some writers use sinc x to mean sin x/x instead of the correct sin $\pi x/\pi x$. (The form sin x/x is unlikely to arise using Woodward's methods, so these writers are mixing the standard integration method with Woodward's misrepresented notation, leading to inconsistency.) The comb function and the rep operator are used for describing sampled or repetitive waveforms and their spectra, and hence enable the whole field of Fourier series to be incorporated into the field of Fourier transforms. Thus, the Fourier series can now be seen as a particular form of the Fourier transform, rather than the Fourier transform seen just as a limiting case of the Fourier series. For suitable waveforms, this enables the Fourier series coefficients to be obtained without explicit integration.

1.3 Outline of the Rules and Pairs Method

To use the rules and pairs, the function to be transformed must first be expressed carefully in the notation in which they are expressed, that is, in terms of the basic functions included in the table of Fourier transform pairs. This table gives the transforms of these functions and the table of rules provides the relationships between these transforms (sums, products, convolutions, and appropriate scaling factors, for example) as determined by the relationships between the input basic functions.

The notation is specific and specialized, but is reasonably natural and quickly absorbed, and it is given in Chapter 2 with the tables of the rules and pairs. When the transform has been obtained using the rules and pairs, the resulting expression requires interpretation and may benefit from rearranging and simplifying. Sketches of the functions can be useful to bring to life a mathematical expression, and illustrations of functions and their transforms have been provided fairly generously, particularly in Chapters 2 and 3.

A feature of the notation is that a given function (or waveform) can sometimes be correctly described in more than one way, leading to more than one expression for its transform (or spectrum), though of course these expressions must be equivalent. One of these expressions will be more appropriate and convenient for one particular case under study than another, so there is an art, based on experience and imagination (as required for solving certain differential equations or some problems in integration), in choosing the description of the waveform that will yield the spectrum represented in the required form. This alternative description approach is particularly effective in producing general results or theorems, as Woodward [1] shows. A good example is in Woodward's proof of the sampling theorem in the time domain, given in Chapter 4, where, by expressing the spectrum of a waveform in two different ways, the equivalence of a continuous waveform and its sampled form is established.

1.4 The Fourier Transform and Generalized Functions

The concept of the Fourier series seems intuitively very reasonable: that any periodic function can be represented by a sum of elementary periodic functions, either sine and cosine functions or, equivalently, complex exponentials (cisoids). The frequencies of the elementary functions are integer multiples (including zero, giving a constant function) of the repetition frequency of the periodic function. The sum may turn out to be infinite, but users of this mathematical tool are generally content to let mathematicians justify such a sum, determining the conditions under which it converges; however, for problems arising in practice, in physics or engineering, for example, it is "obvious" that such a sum does converge. Thus, we can put for f a real or complex function of a real variable x, with period X [so that $f(x + X) = f(x)$],

$$f(x) = \sum_{n=0}^{\infty} a_n \cos(2\pi nx/X) + \sum_{n=1}^{\infty} b_n \sin(2\pi nx/X) = \sum_{n=-\infty}^{\infty} c_n \exp(2\pi inx/X) \tag{1.1}$$

(By expressing the trigonometric functions as complex exponentials, we can relate c_n to a_n and b_n. From now on we restrict our attention exclusively to the complex exponential series.) The coefficients of the series are found by integration over one cycle of the function, so that, for example,

$$c_n = \frac{1}{X} \int_{x_0 - X/2}^{x_0 + X/2} f(x) \exp(-2\pi inx/X)\, dx \tag{1.2}$$

The Fourier transform can be obtained as the limiting case of the Fourier series when the period is increased towards infinity and the fundamental frequency falls to zero. In this case, as $X \to \infty$ we put $n/X \to y$, $1/X \to dy$, $c_n \to g(y)\, dy$, where g is a continuous function replacing the discrete series c_n and the summations in (1.1) become integrals. Thus (1.1) and (1.2) become, respectively,

$$f(x) = \int_{-\infty}^{\infty} g(y) \exp(2\pi ixy)\, dy \tag{1.3}$$

and

$$g(y) = \int_{-\infty}^{\infty} f(x) \exp(-2\pi ixy)\, dx \tag{1.4}$$

Here $g(y)$ is the Fourier transform of $f(x)$. Even practical users, not concerned about problems of convergence and knowing that they have a well-behaved continuous function, with no poles, for example, and believing that there is a well-behaved solution for their problem, will find there is a difficulty here. Whereas it is clear that the integral in (1.2) converges (has a finite value) because it is over a finite interval, the same does not necessarily hold for the integral in (1.4), which is over an infinite interval. (The former is absolutely integrable for such a function—that is, the integral of the modulus of the integrand is finite—and the latter is not necessarily so.) The simplest function for which this difficulty arises is the constant function, and it is clear that the value of this mathematical tool would be severely limited if it could not handle even this case.

An approach to finding the Fourier transform of a constant function, say $f(x) = 1$ for all real x, is to find a sequence of functions that do have transforms as given by (1.4) and that approach f in the limit of some parameter. For example, we could choose $f_n(x)$ to be the function $\exp(-\pi n^2 x^2)$. Putting this into (1.4), we find that its transform is $g_n(y) = n \exp(-\pi y^2/n^2)$. We see that in the limit as $n \to \infty$, $f_n(x) \to f(x) = 1$, in that however small we choose the positive number ϵ, for any x we can find a value of n such that $f_n(x) > 1 - \epsilon$. Also, as $n \to \infty$ we have $g_n(0) \to \infty$, and for nonzero values of y, $g_n(y) \to 0$, as we can always find a value of n such that $g_n(y) < \epsilon$, for any positive ϵ. The limiting function g is the Dirac δ-function, which plays an important part in this theory. It is not strictly a function in the ordinary sense, but is a generalized function in Lighthill's terminology [2].

The fact that there is some difficulty in using (1.4) where $f(x)$ is a constant function (and also if it is periodic) does not mean that such functions do not have Fourier transforms. This problem has been tackled formally and the subject of Fourier transforms put on a rigorous basis by Laurent Schwarz. However, to do this it was necessary to generalize the idea of a function to include the δ-function, and indeed the term *generalized function* has been introduced by Temple and presented clearly and accessibly by Lighthill [2]. As it is shown in this text that, in general, ordinary functions satisfy the definition of generalized functions (as a limiting sequence of suitable functions), the practical result is that we can confidently accept and include δ-functions (and rows of δ-functions, in the case of a line spectrum) with ordinary functions for the purpose of Fourier transform operations and analysis.

The δ-function has been obtained as the limit of a series of Gaussian functions above (also in Woodward [1, p. 28]), the limit of a series of triangular functions [1, p. 15], and the limit of a series of rectangular functions (Figure 2.3). The fact that different sequences can be used is included in Lighthill's definition, though he adds the condition that the functions should be differentiable everywhere, which actually rules out the series of triangular and rectangular functions.

There is no reference to generalized function theory by Woodward; Schwarz's work was published in 1950–1951, only shortly before Woodward's book (1953), and the further spreading of these ideas by Temple (1955) and Lighthill (1958) followed later. The Dirac δ-function is an example of the not uncommon case where physics and engineering have required a new mathematical tool. This has been devised and given a very reasonable justification (for example, as the limit of a series of triangular functions), only later to be given a more rigorous mathematical definition.

1.5 Complex Waveforms and Spectra in Signal Processing

The approach to Fourier transforms presented here uses the complex Fourier transform, by which a waveform, real or complex, is expressed as a sum or integral of complex exponentials [see (1.3)], which are elementary complex waveforms. The idea of a complex waveform should not be seen as just a mathematical convenience, with the "real-world" waveform just the real part. The elementary complex waveform $\exp(2\pi ift)$ can be represented as the pair of real waveforms $\cos(2\pi ft)$ and $\sin(2\pi ft)$ in two channels, which must be handled appropriately—that is, according to the rules of complex arithmetic. We recall that a complex number can be represented as an ordered pair of real numbers; that is, $z = x + iy$ can be written as (x, y), satisfying the rules $(x_1, y_1) + (x_2, y_2) = (x_1 + x_2, y_1 + y_2)$ and $(x_1, y_1) \cdot (x_2, y_2) = (x_1x_2 - y_1y_2, x_1y_2 + x_2y_1)$. This avoids the explicit use of the imaginary constant i, if this worries the practical user, but we need only consider that i acts as a form of switch that moves a waveform from channel 1 (real) to channel 2 (imaginary), or from channel 2 to channel 1, with a sign change in this case. We note that by using complex waveforms, meaning is given to the idea of negative frequency. Compared with the positive frequency form, this corresponds to an inversion of the waveform in the second channel.

In signal processing, it is convenient to use the analytic signal, which is the complex waveform corresponding to the real waveform that is present, as received, for example, from a radio or radar antenna or sonar sensor. Thus if the waveform is expressed as $a(t)\cos[2\pi f_0 t + \phi(t)]$, that is, a carrier at intermediate frequency (IF) or radio frequency (RF) f_0, with amplitude a and phase ϕ, which may be time-varying, in the general case, then we form the complex form $a(t)\exp i[2\pi f_0 t + \phi(t)]$, which is the pair $\{a(t)\cos[2\pi f_0 t + \phi(t)], a(t)\sin[2\pi f_0 t + \phi(t)\}$. The second member of this pair is obtained from the first by a Hilbert transform, which in effect performs a wideband −90-degree phase shift. (Thus all cosine components in the signal, whatever their frequencies, become sines, and sines become −cosines.) In practice, this can be achieved, with a high degree of fidelity (for moderate fractional bandwidths) by a 3-dB hybrid directional coupler. The two (real) outputs of this coupler can be considered as the required (complex) waveform pair. The advantage of using the analytic signal is that (at least when on a carrier) the spectrum is "one-sided," with only positive frequency components, in contrast to the two-sided spectrum of the real waveform. If the complex waveform is then mixed down to complex baseband, using the complex local oscillator (LO) $\exp(-2\pi if_0 t)$ we obtain the complex waveform $a(t)\exp[i\phi(t)]$, which is the part of the waveform, the modulation, containing the information of interest. At baseband the spectrum will contain both positive and negative frequencies from the components of the incoming signal with frequencies above and below the LO frequency, respectively. The spectrum is not necessarily symmetrical in general.

If the two real baseband waveforms are sampled (simultaneously) and digitized, the pairs of samples are available as complex numbers for any processing computation required. The two signal channels are commonly referred to as I and Q, for *i*n-phase and in-*q*uadrature, with respect to the IF form. This is rather clumsy, as the I refers to the real channel, rather than the imaginary. A neater terminology would be P and Q for in-*p*hase and in-*q*uadrature.

1.6 Outline of the Contents

The rules and pairs themselves are presented in Chapter 2, but before they are given, the notation in which they are expressed is defined and illustrated first. Three examples using the technique are then given, which provide an introduction to the method and show how easily some useful and important results can be obtained. Two appendices of parenthetical material are added; one gives an outline of the derivations of the rules and pairs and the other obtains the properties of the extremely useful sinc function, the transform of the rectangular pulse, using the rules.

All the following chapters provide examples and illustrations of the use of the technique. It will be seen that all the results have been obtained without any (explicit) integration whatsoever, and indeed, except for some expressions that are used to define the problem in terms of Fourier transforms, there are no symbols of integration to be seen. The first of these, Chapter 3, on pulse spectra, covers one of the most natural applications of the technique. For readers new to the method, Chapters 2 and 3 should provide a relatively straightforward introduction to its use.

Sampling, particularly relevant for signal processing, is studied in Chapter 4. The basic sampling theorems are given (following Woodward's examples [1]) which give the minimum sampling rate necessary to retain all the information in a waveform of finite bandwidth. Some further forms of sampling are also analyzed, which may be of more theoretical than practical interest. These results are certainly obtained more easily than in the earlier papers on these sampling methods, which did not use Woodward's method and notation.

The question of deriving a series of samples offset in time from the original series is considered in Chapter 5. These interpolated samples correspond to the samples that would have been obtained by sampling the waveform with the time offset. The ability to do this, when the waveform is no longer available, is important, as it provides a sampled form of the delayed waveform. If the waveform is sampled at the minimum rate to retain all the waveform information, accurate interpolation requires combining a substantial number of input samples for each output value. It is shown that oversampling—sampling at a higher rate than actually necessary—can reduce this number very considerably, to quite a low value. The user can compare the disadvantage (if any) of sampling slightly faster with the saving on the amount of computation needed for the interpolation. One example [from a simulation of a radar moving target indication (MTI) system] is given where the reduction in computation can be very great indeed.

The problem of compensating for spectral distortion is considered in Chapter 6. Compensation for delay (a phase error that is linear with frequency) is achieved by a technique similar to interpolation, but amplitude compensation is interesting in that it requires a new set of transform pairs, including functions derived by differentiation of the sinc function. The compensation is seen to be very effective for the problems chosen, and again oversampling can greatly reduce the complexity of the implementation. The problem of equalizing the response of a wideband antenna array used for a radar application is used as an illustration, giving some impressive results.

Finally, in Chapter 7 we take advantage of the fact that there is a Fourier transform relationship between the illumination of a linear aperture and its beam pattern. In fact, rather than a continuous aperture, we concentrate on the regular linear array, which is a sampled aperture and mathematically has a correspondence with the sampled waveforms considered in earlier chapters. Two forms of the problem are considered: the low side-lobe directional beam and a much wider sector beam, covering an angular sector with uniform gain. Similar results could be achieved, in principle, for the continuous aperture, but it would be difficult in practice to apply the required aperture weighting (or tapering).

We note that Chapters 4 through 7 and some of Chapter 3 analyze periodic waveforms (with line spectra) or sampled waveforms (with periodic spectra), implying a requirement for Fourier series analysis rather than the nonperiodic Fourier transform. However, it would not make the problems any easier to turn to conventional Fourier series analysis. As stated earlier, the classical Fourier series theory is now, as Lighthill remarks [2, p. 66], included in the more general Fourier transform approach. Using Woodward's notation, the ease with which the method applies to nonperiodic functions applies also to periodic ones, and no distinction, except in notation, is needed.

References

[1] Woodward, P. M., *Probability and Information Theory, with Applications to Radar,* London: Pergamon Press, 1953; reprint Norwood, MA: Artech House, 1980.

[2] Lighthill, M. J., *Fourier Analysis and Generalised Functions,* Cambridge, UK: Cambridge University Press, 1958.

2

Rules and Pairs

2.1 Introduction

In this chapter we present the basic tools and techniques for carrying out Fourier transforms of suitable functions without using integration. In the rest of the book the definitions and results given here will be used to obtain useful results relatively quickly and easily. Some of these results are well established, but these derivations will serve as valuable illustrations of the method, indicating how similar or related problems may be tackled.

The method has already been outlined in Chapter 1. First, the function to be transformed is described formally in a suitable and precise notation. This defines the function in terms of some very basic, or elementary, functions, such as rectangular pulses or δ-functions, which are combined in various ways, such as by addition, multiplication, or convolution. Each of these elementary functions has a Fourier transform, the function and its transform forming a transform pair. Next, the transform is carried out by using the known set of pairs to replace each elementary waveform with its transform, and also by using a set of established rules that relates the way the transforms are combined to the way the input functions were combined. For example, addition, multiplication, and convolution of functions transform to addition, convolution, and multiplication of transforms, respectively. Finally, the transform expression needs interpretation, possibly after rearrangement. Diagrams of the functions and transforms can be helpful and are widely used here.

We begin by defining the notation used. Some of these terms, such as rect and sinc, have been adopted more widely to some extent, but rep and comb are less well known. We include a short discussion on convolution, as this operation is important in this work, being the operation in the transform domain corresponding to multiplication in the original domain (and vice versa). This is followed by the rules relating to Fourier transforms and a set of Fourier transform pairs. We then include three illustrations as examples before the main applications in the following chapters.

2.2 Notation

2.2.1 Fourier Transform and Inverse Fourier Transform

Let u and U be two (generalized) functions related by

$$u(x) = \int_{-\infty}^{\infty} U(y)e^{2\pi ixy}\,dy \tag{2.1}$$

and

$$U(y) = \int_{-\infty}^{\infty} u(x)e^{-2\pi ixy}\,dx \tag{2.2}$$

U is the Fourier transform of u, and u is the inverse Fourier transform of U. We have used a general pair of variables, x and y, for the two transform domains, but in the very widespread application of these transforms in spectral analysis of time-dependent waveforms, we choose t and f, associated with time and frequency. We take the transforms in this form, with 2π in the exponential (so that in spectral analysis, for example, we use the frequency f, rather than the angular frequency $\omega = 2\pi f$) in order to maintain a high degree of symmetry between the variables; otherwise we need to introduce a factor of $1/2\pi$ in one of the expressions, for the transform, or $1/\sqrt{2\pi}$ in both. We find it convenient to keep generally to a convention of using lower case letters for the waveforms, or primary domain functions, and upper case for their transforms, or spectra. We indicate a Fourier transform pair of this kind by

$$u \Leftrightarrow U \tag{2.3}$$

with $\Rightarrow$ implying the forward transform and $\Leftarrow$ the inverse.

We note that there remains a small asymmetry between the expressions; the forward transform has a negative exponent and the inverse has a positive exponent. Many functions used are symmetric and for these the forward and inverse transform operations are identical. However, when this is not the case, it may be important to note just which transform is needed in a given application.

2.2.2 rect and sinc

The rect function is defined by

$$\operatorname{rect} x = \begin{cases} 1 & \text{for } -\dfrac{1}{2} < x < +\dfrac{1}{2} \\ 0 & x < -\dfrac{1}{2} \text{ and } x > +\dfrac{1}{2} \end{cases} \qquad (x \in \mathbb{R}) \tag{2.4}$$

[and rect $(\pm\frac{1}{2}) = \frac{1}{2}$]. This is a very commonly encountered gating function. This pulse is of unit width, unit height and is centered at zero [Figure 2.1(a)]. A pulse of width T, amplitude A and centered at time t_0 is given by $A\operatorname{rect}[(t - t_0)/T]$, shown in Figure 2.1(b). In the frequency domain, a rectangular frequency band of width B, centered at f_0, is defined by $\operatorname{rect}[(f - f_0)/B]$. A pulse, or a filter, with this characteristic is not strictly realistic (or realizable) but may be sufficiently close for many investigations.

The Fourier transform of the rect function is the sinc function, given by

$$\operatorname{sinc} x = \begin{cases} \sin(\pi x)/\pi x & \text{for } x \neq 0 \\ 1 & \text{for } x = 0 \end{cases} \qquad (x \in \mathbb{R}) \tag{2.5}$$

This is illustrated in Figure 2.2(a), and a shifted, scaled form is shown in Figure 2.2(b). This follows Woodward's definition [1] and is a neater

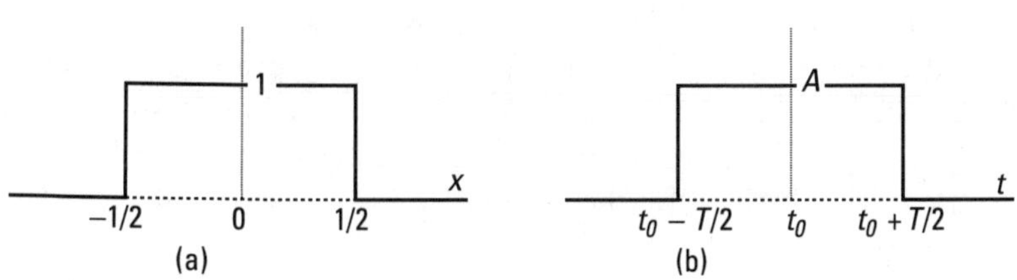

Figure 2.1 rect functions: (a) rect(x); (b) $A\operatorname{rect}[(t - t_0)/T]$.

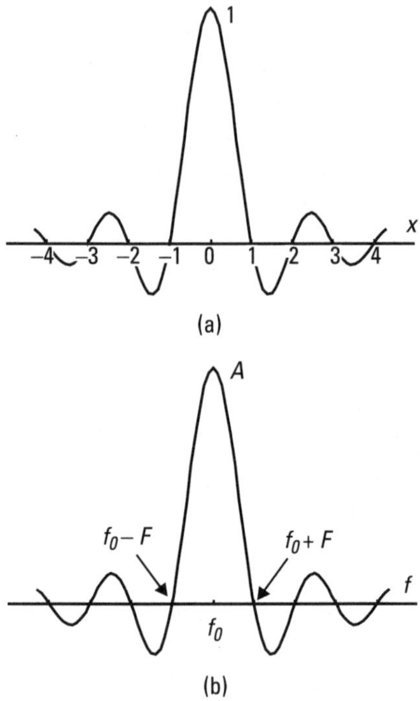

Figure 2.2 sinc functions: (a) sinc (x); (b) A sinc $[(f-f_0)/F]$.

function than sin x/x, which is sometimes (wrongly) called sinc x. It has the following properties:

1. sinc $n = 0$, for n a nonzero integer
2. $\int_{-\infty}^{\infty} \text{sinc}\, x\, dx = 1$
3. $\int_{-\infty}^{\infty} \text{sinc}^2 x\, dx = 1$
4. $\int_{-\infty}^{\infty} \text{sinc}\,(x-m)\, \text{sinc}\,(x-n)\, dx = \delta_{mn}$

where m and n are integers and δ_{mn} is the Kronecker-δ. (For the function sin x/x, the results are more untidy, with π or π^2 appearing.) The last two results can be stated in the following form: the set of shifted sinc functions $\{\text{sinc}\,(x-n): n \in \mathbb{Z}, x \in \mathbb{R}\}$ is an orthonormal set on the real line. These results are easily obtained by the methods presented here, and are derived in Appendix 2A.

Despite the $1/x$ factor, this function is analytic on the real line. The only point where this property may be in question is at $x = 0$. However, as

$$\lim_{x \to +0} \operatorname{sinc} x = \lim_{x \to -0} \operatorname{sinc} x = 1$$

by defining sinc (0) = 1, we ensure that the function is continuous and differentiable at this point. Useful facts about the sinc function are that its 4-dB beamwidth is almost exactly equal to half the width at the first zeros [± 1 in the basic function and $\pm F$ in the scaled version of Figure 2.2(b)], the 3-dB width is 0.886, and the first side-lobe peak is at the rather high level of -13.3 dB relative to the peak of the main lobe.

2.2.3 δ-Function and Step Function

The δ function is not a proper function but can be defined as the limit of a sequence of functions that have integral unity and that converge pointwise to zero everywhere on the real line except at zero. [Suitable sequences of functions f_n such that $\lim_{n \to \infty} f_n(x) = \delta(x)$ are n rect nx and $n \exp(-2\pi n^2 x^2)$, illustrated in Figure 2.3.] This function consequently has the properties

$$\delta(x) = \begin{cases} \infty & \text{for } x = 0 \\ 0 & \text{for } x \neq 0 \end{cases} \qquad (x \in \mathbb{R}) \tag{2.6}$$

$$\int_{-\infty}^{\infty} \delta(x)\, dx = 1 \tag{2.7}$$

(In fact, the generalized function defined by Lighthill [2] requires the members of the sequence to be differentiable everywhere; this actually rules out the rect function sequence.) From (2.6) and (2.7) we deduce the important property that

$$\int_{\mathrm{I}} \delta(x - x_0)\, u(x)\, dx = u(x_0) \tag{2.8}$$

as the integrand is zero everywhere except at x_0, and I is any interval containing x_0. Thus the convolution (defined below) of a function u with a δ-function at x_0 is given by

$$u(x) \otimes \delta(x - x_0) = \int_{-\infty}^{\infty} u(x - x')\, \delta(x' - x_0)\, dx' = u(x - x_0) \tag{2.9}$$

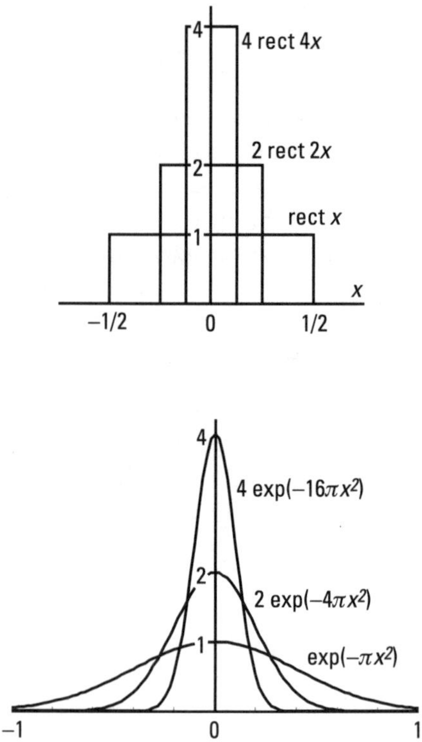

Figure 2.3 Two series approximating δ-functions.

That is, the waveform is shifted so that its previous origin becomes the point x_0, the position of the δ-function. The function u itself could be a δ-function; for example,

$$\delta(x-x_1) \otimes \delta(x-x_2) = \int_{-\infty}^{\infty} \delta(x-x'-x_1)\,\delta(x'-x_2)\,dx' = \delta[x-(x_1+x_2)] \tag{2.10}$$

Thus, convolving δ-functions displaced by x_1 and x_2 from the origin gives a δ-function at $(x_1 + x_2)$.

The δ-function in the time domain represents a unit impulse occurring at the time when the argument of the δ-function is zero, that is, $\delta(t - t_0)$ represents a unit impulse at time t_0. In the frequency domain, it represents a spectral line of unit power. A scaled δ-function, such as $A\delta(x - x_0)$, is described as being of strength A. In diagrams, such as Figure 2.6 below, it is represented by a vertical line of height A at position x_0.

The unit step function $h(x)$, shown in Figure 2.4(a), is here defined by

$$h(x) = \begin{cases} 1 & \text{for } x > 0 \\ 0 & \text{for } x < 0 \end{cases} \qquad (x \in \mathbb{R}) \tag{2.11}$$

[and $h(0) = ½$]. It can also be defined as the integral of the δ-function:

$$h(x) = \int_{-\infty}^{x} \delta(\xi)\, d\xi \tag{2.12}$$

and the δ-function is the derivative of the step function.

The step function with the step at x_0 is given by $h(x - x_0)$ [Figure 2.4(b)].

2.2.4 rep and comb

The rep operator produces a new function by repeating a function at regular intervals specified by its suffix. For example, if $p(t)$ is a description of a pulse, an infinite sequence of pulses at the repetition interval T is given by $u(t)$, shown in Figure 2.5, where

$$u(t) = \mathrm{rep}_T\, p(t) = \sum_{n=-\infty}^{\infty} p(t - nT) \tag{2.13}$$

The shifted waveforms $p(t - nT)$ may be overlapping. This will be the case if the duration of p is greater than the repetition interval T. Any

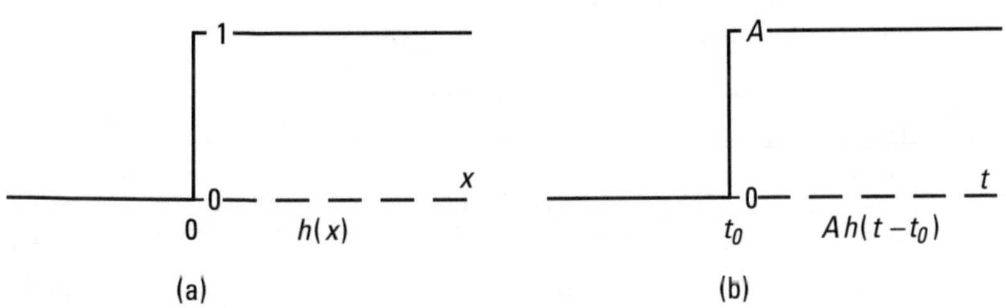

Figure 2.4 Step functions: (a) the unit step; (b) a scaled and shifted step.

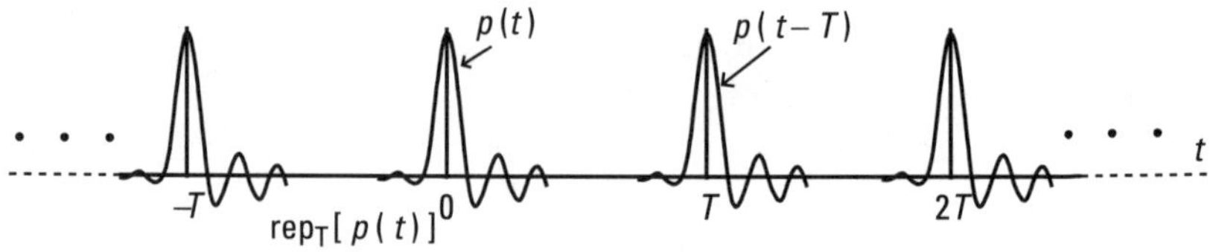

Figure 2.5 The rep operator.

repetitive waveform can be expressed as a rep function—any section of the waveform one period long can be taken as the basic function, and this is then repeated (without overlapping) at intervals of the period.

The comb operator applied to a continuous function replaces the function with δ-functions at regular intervals, specified by the suffix, with strengths given by the function values at those points, that is,

$$\operatorname{comb}_T u(t) = \sum_{n=-\infty}^{\infty} u(nT)\,\delta(t-nT) \tag{2.14}$$

In the time domain this represents an ideal sampling operation. In the frequency domain the comb version of a continuous spectrum is the line spectrum corresponding to the repetitive form of the waveform that gave the continuous spectrum.

The function $\operatorname{comb}_T u(t)$ is illustrated in Figure 2.6, where $u(t)$ is the underlying continuous function, shown dotted, and the comb function is the set of δ-functions.

2.2.5 Convolution

We denote the convolution of two functions u and v by $\otimes$, so that

$$u(x) \otimes v(x) = \int_{-\infty}^{\infty} u(x-x')\,v(x')\,dx' = \int_{-\infty}^{\infty} u(x')\,v(x-x')\,dx' \tag{2.15}$$

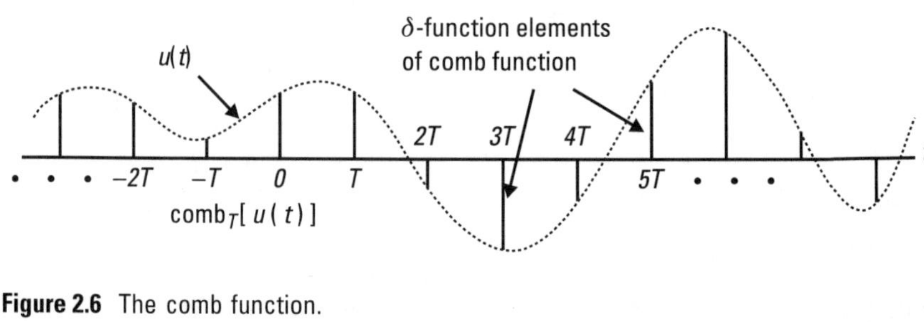

Figure 2.6 The comb function.

One reason for requiring such a function is to find the response of a linear system to an input $u(t)$ when the system's response to a unit impulse (at time zero) is $v(t)$. The response at time t to an impulse at time t' is thus $v(t-t')$. We divide u into an infinite sum of impulses $u(t')\,dt'$ and integrate, so that the output at time t is

$$\int_{-\infty}^{\infty} u(t')\,v(t-t')\,dt' = u(t) \otimes v(t) \tag{2.16}$$

The reason for the reversal of the response v (as a function of t') is because the *later* the impulse $u(t')\,dt'$ arrives, the *earlier* in the impulse response is its contribution to the total response at time t.

It is clear, from the linear property of integration, that convolution is distributive and linear so that we have

$$u \otimes (av + bw) = au \otimes v + bu \otimes w \tag{2.17}$$

where a and b are constants. It is also the case that convolution is commutative (so $u \otimes v = v \otimes u$) and associative, so that

$$u \otimes (v \otimes w) = (u \otimes v) \otimes w \tag{2.18}$$

and we can write these simply as $u \otimes v \otimes w$ without ambiguity. Thus we are free to rearrange combinations of convolutions within these rules and evaluate multiple convolutions in different sequences, as shown in (2.18).

It is useful to have a feel for the meaning of the convolution of two functions. The convolution is obtained by sliding one of the functions (reversed) past the other and integrating the point-by-point product of the functions over the whole real line. Figure 2.7(a) shows the result of convolving two rect functions, $\text{rect}(t/T_1)$ and $\text{rect}(t/T_2)$, with $T_1 < T_2$, and Figure 2.7(b) shows that the value of the convolution at point $-t_0$ is given by the area of overlap of the functions when the "sliding" function, $\text{rect}(t/T_1)$, shown dashed, is centered at $-t_0$. We note that overlap begins when $t = -(T_1 + T_2)/2$, and increases linearly until the smaller pulse is within the larger, at $-(T_1 - T_2)/2$. The magnitude of the flat top is just T_1, the area of the smaller pulse, for these unit height pulses. This is equal to the area of overlap when the narrower pulse is entirely within the wider one. For pulses of magnitudes A_1 and A_2, the level would be $A_1 A_2 T_1$, and for pulses centered at t_1 and t_2, the convolved response would be centered at $t_1 + t_2$.

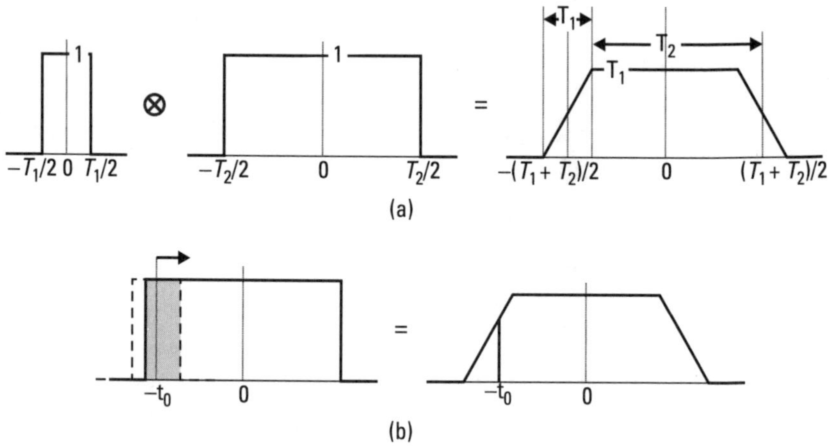

Figure 2.7 Convolution of two rect functions: (a) full convolution; (b) value at a single point.

In many cases we will be convolving symmetrical functions such as rect or sinc, but if we have a nonsymmetric one it is important to note from (2.15) that $u(x - x')$, considered as a function of x', is not only shifted by x (the sliding parameter), but is reversed with respect to $u(x')$. In Figure 2.8(a) we show the result of convolving an asymmetric triangular pulse with a rect function, and in Figure 2.8(b) we show, on the left, that the reversed triangular pulse is used when it is the sliding function; on the right we show that, because of the commutativity of convolution, we could equally well

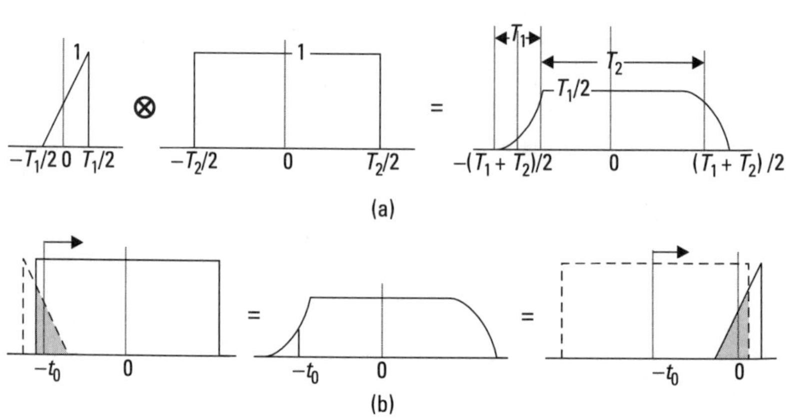

Figure 2.8 Convolution with a nonsymmetric function: (a) full convolution; (b) value at a single point.

use the rect function as the moving one, which, being symmetric, is unchanged when reversed, of course.

2.3 Rules and Pairs

The rules and pairs at the heart of this technique of Fourier analysis are given in Tables 2.1 and 2.2 below. The rules are relationships that apply generally to all functions (u and v in Table 2.1) and their transforms (U

Table 2.1
Rules for Fourier Transforms

Rule	Function	Transform	Notes
—	$u(x)$	$U(y)$	See (2.1), (2.2)
1	$au + bv$	$aU + bV$	a, b constants ($a, b \in \mathbb{C}$, in general)
2	$u(-x)$	$U(-y)$	
3	$u^*(x)$	$U^*(-y)$	* indicates complex conjugate
4	$U(x)$	$u(-y)$	
5	$u(x/X)$	$\|X\| U(Xy)$	$X \in \mathbb{R}$, X constant
6a	$u(x - x_0)$	$U(y) \exp(-2\pi i x_0 y)$	$x_0 \in \mathbb{R}$, x_0 constant
6b	$u(x) \exp(2\pi i x y_0)$	$U(y - y_0)$	$y_0 \in \mathbb{R}$, y_0 constant
7a	uv	$U \otimes V$	(2.15)
7b	$u \otimes v$	UV	
8a	$\text{comb}_X u$	$\|Y\| \text{rep}_Y U$	(2.14), (2.13), $Y = 1/X$, constant
8b	$\text{rep}_X u$	$\|Y\| \text{comb}_Y U$	
9a	$u'(x)$	$2\pi i y U(y)$	Prime indicates differentiation
9b	$-2\pi i x u(x)$	$U'(y)$	
10a	$\int_{-\infty}^{x} u(\xi)\, d\xi$	$U(y)\left[\frac{\delta(y)}{2} + \frac{1}{2\pi i y}\right]$	
10b	$u(x)\left[\frac{\delta(x)}{2} - \frac{1}{2\pi i x}\right]$	$\int_{-\infty}^{y} U(\eta)\, d\eta$	

Table 2.2
Fourier Transforms Pairs

Pair	Function	Transform	Notes
1a	$\delta(x)$	1	(2.6)
1b	1	$\delta(y)$	
2a	$h(x)$	$\frac{\delta(y)}{2} + \frac{1}{2\pi iy}$	(2.11)
2b	$\frac{\delta(x)}{2} - \frac{1}{2\pi ix}$	$h(y)$	
3a	$\operatorname{rect}(x)$	$\operatorname{sinc}(y)$	(2.4), (2.5)
3b	$\operatorname{sinc}(x)$	$\operatorname{rect}(y)$	
4	$\exp(-x)$	$\frac{1}{1 + 2\pi iy}$	$(x \geq 0)$ Laplace transform
5	$\exp(-\pi x^2)$	$\exp(-\pi y^2)$	
6a	$\delta(x - x_0)$	$\exp(-2\pi ix_0 y)$	P1a, R6a
6b	$\exp(2\pi iy_0 x)$	$\delta(y - y_0)$	P1b, R6b
7a	$\cos 2\pi y_0 x$	$(\delta(y - y_0) + \delta(y + y_0))/2$	P6b, R1
7b	$\sin 2\pi y_0 x$	$(\delta(y - y_0) - \delta(y + y_0))/2i$	P6b, R1
8a	$u(x)\cos 2\pi y_0 x$	$(U(y - y_0) + U(y + y_0))/2$	P7a, R7a, (2.17)
8b	$u(x)\sin 2\pi y_0 x$	$(U(y - y_0) - U(y + y_0))/2i$	P7a, R7a, (2.17)
9	$\exp(-ax)$	$1/(a + 2\pi iy)$	$(a > 0,\ x \geq 0)$ P4, R5
10	$\exp(-x^2/2\sigma^2)$	$\sigma\sqrt{2\pi}\exp(-2\pi^2\sigma^2 y^2)$	P5, R5
11	$\operatorname{comb}_X(1)$	$\lvert Y\rvert \operatorname{comb}_Y(1)$	$Y = 1/X$

a, x_0, y_0, X, Y, σ all real constants and also $x, y \in \mathbb{R}$

and V). The pairs are certain specific Fourier transform pairs. All these results are proved, or derived in outline, in Appendix 2B.

In Table 2.1 the rules labeled "b" are derivable from those labeled "a," using other rules, but it is convenient for the user to have both a and b versions. We see that there is a great deal of symmetry between the a and b versions, with differences of sign in some cases.

To illustrate such a derivation, we derive Rule 6b from Rule 6a. Let U be a function of x with transform V; then from Rule 6a,

$$U(x - x_0) \Leftrightarrow V(y)\exp(-2\pi ix_0 y)$$

From Rule 4, if $u(x) \Leftrightarrow U(y)$, then $U(x) \Leftrightarrow u(-y)$, so in this case we have

$$U(x) \Leftrightarrow V(y) = u(-y)$$

and so

$$U(x - x_0) \Leftrightarrow u(-y)\exp(-2\pi i x_0 y) \tag{2.19}$$

Now we use Rule 4 again, in reverse; that is, if $Z(x) \Leftrightarrow z(-y)$, then $z(x) \Leftrightarrow Z(y)$, so that (2.19) becomes

$$u(x)\exp(2\pi i y_0 x) \Leftrightarrow U(y - y_0)$$

(on renaming the constant x_0 as y_0), and this is Rule 6b. However, in this case, the result is easily obtained from the definitions of the Fourier transform in (2.2), as shown in Appendix 2B.

In Table 2.2, not only are pairs 1b, 2b, and 3b derivable from the corresponding a form, but the pairs 6 to 10 are all derivable from other pairs using the rules, and these are indicated by the P and R notation, which will be used subsequently. Although they are not fundamental, these results are included for convenience, as they occur frequently.

An important point follows from Rule 3. For a real waveform, we have

$$u(t) = u(t)^*$$

so, from R3,

$$U(f) = U(-f)^* \tag{2.20}$$

or

$$U_R(f) + iU_I(f) = U_R(-f) - iU_I(-f) \tag{2.21}$$

where U_R and U_I are the real and imaginary parts of U.

We see from (2.20) that for a real waveform the negative frequency part of the spectrum is simply the complex conjugate of the positive frequency part and contains no extra information. It follows [see (2.21)] that the real part of the spectrum of a real function is always an even function of frequency and the imaginary part is an odd function. (Often spectra of simple waveforms are either purely real or imaginary—see P7a and P7b above, for example). Thus, for real waveforms, we need only consider the positive frequency part of the spectrum, remembering that the power at a given frequency is twice the power given by this part, because there is an equal contribution from the negative frequency component. (A short discussion and interpretation of negative frequencies was given in Section 1.5 above.)

2.4 Three Illustrations

2.4.1 Narrowband Waveforms

The case of waveforms modulated on a carrier is described by P8a or P8b (which could be considered rules as much as pairs). Although these relations apply generally, we consider the frequently encountered narrowband case, where the modulating or gating waveform u has a bandwidth that is small compared with the carrier frequency f_0. We see that the spectrum, in this case, consists of two essentially distinct parts—the spectral function U, centered at f_0 and at $-f_0$. Again, for a real waveform, the negative frequency part of the waveform contains no extra information and can safely be neglected (apart from the factor of two when evaluating powers). However, strictly speaking, the function U centered at $-f_0$ may have a tail that stretches into the positive frequency region, and in particular it may stretch to the region around f_0 if the waveform is not sufficiently narrowband. In that case the contribution of $U(f + f_0)$ in the positive frequency range must not be neglected.

Figure 2.9 shows how the spectrum $U(f)$ of the baseband waveform $u(t)$ is centered at frequencies $+f_0$ and $-f_0$ when modulating (or multiplying) a carrier. When applied to the carrier 2 cos $2\pi f_0 t$, we see, from P7a, that we just have U shifted to these frequencies. When applied to 2 sin $2\pi f_0 t$, we obtain, from P7b, $-iU$ centered at f_0 and iU at $-f_0$. We have chosen a real baseband waveform $u(t)$ so that its spectrum is shown with a symmetric, or even, real part and an antisymmetric, or odd, imaginary part, as shown above for real waveforms. We see that this property holds for the spectrum of the real waveforms $u(t)$ cos $2\pi f_0 t$ and $u(t)$ sin $2\pi f_0 t$.

2.4.2 Parseval's Theorem

Another result, Parseval's theorem, follows easily from the rules. Writing out Rule 7 using the definitions of Fourier transform and convolution [(2.1) and (2.15)] gives

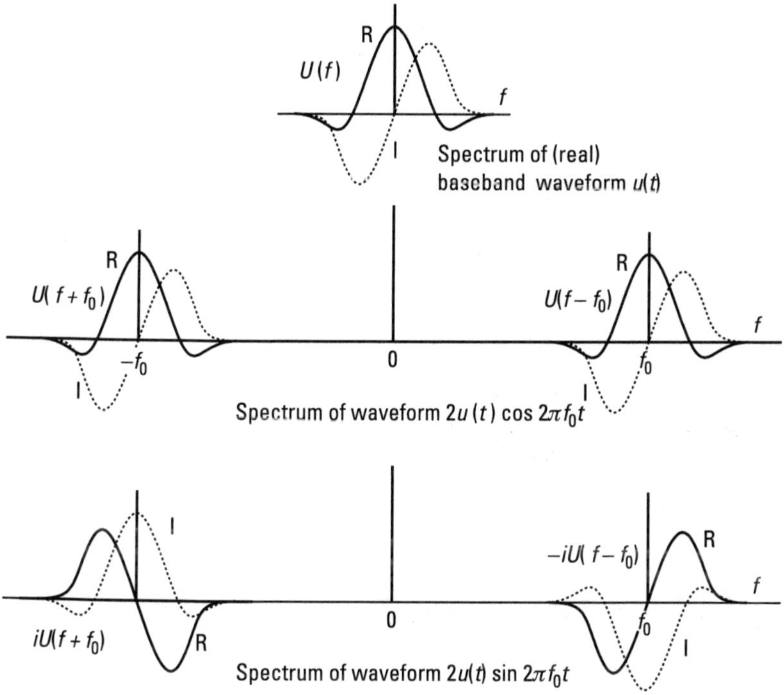

Figure 2.9 Spectra of modulated carrier, (real) narrowband waveforms.

$$\int_{-\infty}^{\infty} u(x)\, v(x)\, e^{2\pi ixy}\, dx = \int_{-\infty}^{\infty} U(\psi)\, V(y - \psi)\, d\psi \tag{2.22}$$

Putting $y = 0$ in this equation and then replacing the variable of integration ψ with y gives

$$\int_{-\infty}^{\infty} u(x)\, v(x)\, dx = \int_{-\infty}^{\infty} U(y)\, V(-y)\, dy \tag{2.23}$$

Replacing v with v^* and using R3 gives Parseval's theorem:

$$\int_{-\infty}^{\infty} u(x)\, v(x)^*\, dx = \int_{-\infty}^{\infty} U(y)\, V(y)^*\, dy \tag{2.24}$$

Taking the particular case of $v = u$ then gives

$$\int_{-\infty}^{\infty} |u(x)|^2\, dx = \int_{-\infty}^{\infty} |U(y)|^2\, dy \tag{2.25}$$

This simply states that the total energy in a waveform is equal to the total energy in its spectrum. For a real waveform we have

$$\int_{-\infty}^{\infty} u(x)^2\, dx = 2\int_{0}^{\infty} |U(y)|^2\, dy \tag{2.26}$$

using $U(y) = U(-y)^*$ for the spectrum of a real waveform.

2.4.3 The Wiener-Khinchine Relation

This states that the autocorrelation function of a waveform is given by the (inverse) Fourier transform of its power spectrum. For a waveform u with (amplitude) spectrum U, the power spectrum is $|U|^2$, and from R2 and R3 we see that $U^*(f)$ is the transform of $u^*(-t)$, so we have

$$u(t) \otimes u^*(-t) \Leftrightarrow U(f) \cdot U^*(f) = |U(f)|^2 \tag{2.27}$$

Writing out the convolution, we have

$$u(t) \otimes u^*(-t) = \int_{-\infty}^{\infty} u(t - t')\, u^*(-t')\, dt' = \int_{-\infty}^{\infty} u(s)\, u(s - t)\, ds = r(t) \tag{2.28}$$

where $s = t - t'$ and $r(t)$ is the autocorrelation function for a delay of t. The delay, or time shift between the correlating waveforms, is generally given the symbol τ, rather than t, used for the usual time variable. Thus we have, from (2.27) and (2.28),

$$r(\tau) \Leftrightarrow |U(f)|^2 \tag{2.29}$$

which is the Wiener-Khinchine relation, obtained very concisely by this method.

References

[1] Woodward, P. M., *Probability and Information Theory, with Applications to Radar,* Norwood, MA: 1980.

[2] Lighthill, M. J., *Fourier Analysis and Generalised Functions,* Cambridge, UK: Cambridge University Press, 1958.

Appendix 2A: Properties of the sinc Function

1. sinc $n = 0$ (n a nonzero integer).

   When $n \neq 0$, as $\sin n\pi = 0$, we have $\operatorname{sinc} n = \sin n\pi / n\pi = 0$.

2. $\int_{-\infty}^{\infty} \operatorname{sinc} x \, dx = 1$

   We can write

$$\int_{-\infty}^{\infty} \operatorname{sinc} x \, dx = \int_{-\infty}^{\infty} \operatorname{sinc} x e^{2\pi i x y} \, dx \Big|_{y=0} = \operatorname{rect} y \big|_{y=0} = 1$$

   Here we have converted the integral into an inverse Fourier transform (though the variable in the transform domain here has the value zero) and used P3.

3. $\int_{-\infty}^{\infty} \operatorname{sinc}^2 x \, dx = 1$

   We have

$$\int_{-\infty}^{\infty} \operatorname{sinc}^2 x \, dx = \int_{-\infty}^{\infty} \operatorname{sinc} x \cdot \operatorname{sinc} x e^{2\pi i x y} \, dx \Big|_{y=0} = \operatorname{rect} y \otimes \operatorname{rect} y \big|_{y=0} = 1$$

   rect $y \otimes$ rect y is a triangular function, with peak value 1 at $y = 0$. (This convolution is shown in Figure 3.4, with $A = 1$ and $T = 1$ in this case.)

4. $\int_{-\infty}^{\infty} \operatorname{sinc}(x-m)\operatorname{sinc}(x-n)\,dx = \delta_{mn}$

Using the result in item 3 above, if $m = n$ the integral is

$$\int_{-\infty}^{\infty} \operatorname{sinc}^2(x-n)\,dx = \int_{-\infty}^{\infty} \operatorname{sinc}^2 x\,dx = 1$$

If $m \neq n$, then

$$\int_{-\infty}^{\infty} \operatorname{sinc}(x-m)\operatorname{sinc}(x-n)\,dx$$

$$= \left.\int_{-\infty}^{\infty} \operatorname{sinc}(x-m)\operatorname{sinc}(x-n)\,e^{2\pi ixy}\,dx\right|_{y=0}$$

$$= e^{-2\pi imy}\operatorname{rect}(y) \otimes e^{-2\pi iny}\operatorname{rect}(y)\Big|_{y=0}$$

on using R6a and P3. Forming the convolution integral, this becomes

$$\left.\int_{-\infty}^{\infty} e^{-2\pi iny'}\operatorname{rect}(y')\,e^{-2\pi in(y-y')}\operatorname{rect}(y-y')\,dy'\right|_{y=0}$$

$$= \int_{-\infty}^{\infty} e^{2\pi i(n-m)y'}\operatorname{rect}(y')\operatorname{rect}(-y')\,dy'$$

$$= \int_{-\infty}^{\infty} e^{2\pi i(n-m)y'}\operatorname{rect}(y')\,dy' = \operatorname{sinc}(n-m) = 0$$

on using $\operatorname{rect}(-y') = \operatorname{rect}(y')$, $\operatorname{rect}^2(y') = \operatorname{rect}(y')$, P3, and result 1 above.

Appendix 2B: Brief Derivations of the Rules and Pairs

2B.1 Rules

R1:

This follows from the linearity of integration.

R2:

$$\int_{-\infty}^{\infty} u(-x)\exp(-2\pi ixy)\,dx = \int_{-\infty}^{\infty} u(z)\exp[-2\pi iz(-y)]\,dz$$

$$= U(-y) \qquad (z = -x)$$

R3:

$$\int_{-\infty}^{\infty} u^*(x)\exp(-2\pi ixy)\,dx = \left\{\int_{-\infty}^{\infty} u(x)\exp[2\pi ix(y)]\,dz\right\}^*$$

$$= \left\{\int_{-\infty}^{\infty} u(x)\exp[-2\pi ix(-y)]\,dz\right\}^*$$

$$= U^*(-y)$$

R4:

$$\int_{-\infty}^{\infty} U(x)\exp(-2\pi ixy)\,dx = \int_{-\infty}^{\infty} U(x)\exp[2\pi ix(-y)]\,dz = u(-y)$$

[using the inverse transform, as in (2.1)].

R5:

(a) $X > 0,\ z = x/X = x/|X|$

$$\int_{-\infty}^{\infty} u(x/X)\exp(-2\pi ixy)\,dx = X\int_{-\infty}^{\infty} u(z)\exp(-2\pi izXy)\,dz$$

$$= XU(Xy) = |X|\,U(Xy)$$

(b) $X < 0,\ z = x/X = -x/|X|$

$$\int_{-\infty}^{\infty} u(x/X)\exp(-2\pi ixy)\,dx = -|X|\int_{\infty}^{-\infty} u(z)\exp\left(2\pi iz|X|y\right)dz$$

$$= |X|\int_{-\infty}^{\infty} u(z)\exp\left[-2\pi iz\left(-|X|y\right)\right]dz$$

$$= |X|\,U\left(-|X|y\right) = |X|\,U(Xy)$$

R6a:

$$\int_{-\infty}^{\infty} u(x-x_0)\exp(-2\pi ixy)\,dx = \int_{-\infty}^{\infty} u(z)\exp[-2\pi i(z+x_0)y]\,dz$$

$$= U(y)\exp(-2\pi ix_0y) \qquad (z = x - x_0)$$

R6b:

$$\int_{-\infty}^{\infty} u(x)\exp(2\pi ixy_0)\exp(-2\pi ixy)\,dx = \int_{-\infty}^{\infty} u(x)\exp[-2\pi ix(y-y_0)]\,dx$$

$$= U(y-y_0)$$

R7a:

$$\int_{-\infty}^{\infty} u(x)\, v(x) \exp(-2\pi ixy)\; dx$$

$$= \int_{-\infty}^{\infty}\int_{-\infty}^{\infty} U(z) \exp(2\pi ixz)\, v(x) \exp(-2\pi ixy)\; dx\; dz$$

$$= \int_{-\infty}^{\infty}\int_{-\infty}^{\infty} U(z)\, v(x) \exp[-2\pi ix(y - z)]\; dx\; dz$$

$$= \int_{-\infty}^{\infty} U(z)\, V(y - z)\; dz = U(y) \otimes V(y)$$

R7b:

The transform of $u(x) \otimes v(x)$ is, using $x - z = t$,

$$\int_{-\infty}^{\infty}\int_{-\infty}^{\infty} u(z)\, v(x - z) \exp(-2\pi ixy)\; dx\; dz$$

$$= \int_{-\infty}^{\infty}\int_{-\infty}^{\infty} u(z)\, v(t) \exp[-2\pi i(z + t)y]\; dt\; dz$$

$$= \int_{-\infty}^{\infty}\int_{-\infty}^{\infty} u(z) \exp(-2\pi izy)\, v(t) \exp(-2\pi ity)\; dt\; dz$$

$$= U(y) \int_{-\infty}^{\infty} v(t) \exp(-2\pi ity)\; dt$$

$$= U(y)\, V(y)$$

R8a:

If $v(x) = \mathrm{comb}_X\, u(x) = \sum_{n=-\infty}^{\infty} u(nX)\, \delta(x - nX)$, then the transform is

$$V(y) = \sum_{n=-\infty}^{\infty} u(nX) \exp(-2\pi inXy)$$

This is in the form of a Fourier series, with period $1/X = Y$, and the coefficients are given by integration of V over one period:

$$u(nX) = \frac{1}{Y}\int_0^Y V(y)\exp(2\pi iny/Y)\,dy$$

Also, from the Fourier transform,

$$u(nX) = \int_{-\infty}^{\infty} U(z)\exp(2\pi inXz)\,dz = \sum_{m=-\infty}^{\infty}\int_{-mY}^{(-m+1)Y} U(z)\exp(2\pi inz/Y)\,dz$$

on dividing the range of integration into units of length Y. Putting $y = z + mY$ for each value of m,

$$u(nX) = \int_0^Y \sum_{m=-\infty}^{\infty} U(y - mY)\exp[2\pi in(y - mY)/Y]\,dy$$

$$= \int_0^Y \sum_{m=-\infty}^{\infty} U(y - mY)\exp(2\pi iny/Y)\,dy$$

Comparing the two expressions for $u(nX)$, we see that

$$V(y) = Y\sum_{m=-\infty}^{\infty} U(y - mY) = Y\,\mathrm{rep}_Y U(y)$$

(This is in line with Woodward's comment [1], following his list of rules and pairs, that the transform relationship between comb and rep "can be justified by resorting to a Fourier Series representation." Note: the rule actually uses $|X|$ rather than X; however, from the definitions it is clear that $\mathrm{rep}_{-X}\,u = \mathrm{rep}_X\,u$ and $\mathrm{comb}_{-X}\,u = \mathrm{comb}_X\,u$, so $|X|$ can replace X.)

R8b:

Letting $v(x) = \mathrm{rep}_X u(x) = \sum_{n=-\infty}^{\infty} u(x - nX)$, which is periodic, with period X, we can put

$$v(x) = \sum_{m=-\infty}^{\infty} a_m \exp(2\pi inx/X)$$

and putting $z = x - mX$,

$$a_m = \frac{1}{X}\int_0^X \sum_{m=-\infty}^{\infty} u(x - mX)\exp(-2\pi inx/X)\ dx$$

$$= \frac{1}{X}\sum_{m=-\infty}^{\infty} \int_{-mX}^{(-m+1)X} u(z)\exp[-2\pi in(z + mX)/X]\ dz$$

$$= \frac{1}{X}\int_{-\infty}^{\infty} u(z)\exp(-2\pi inz/X)\ dz = \frac{1}{X}\,U(n/X)$$

Then

$$v(x) = \frac{1}{X}\sum_{m=-\infty}^{\infty} U(n/X)\exp(2\pi inx/X)$$

and

$$V(y) = \frac{1}{X}\sum_{m=-\infty}^{\infty} U(n/X)\,\delta(y - n/X) = \frac{1}{X}\,\mathrm{comb}_{1/X}\,U(y) = Y\,\mathrm{comb}_Y\,U(y)$$

where $Y = 1/X$.

R9a:

$$u(x) = \int_{-\infty}^{\infty} U(y)\exp(2\pi ixy)\ dy$$

$$u'(x) = \int_{-\infty}^{\infty} 2\pi iyU(y)\exp(2\pi ixy)\ dy$$

so $u'(x)$ is the inverse Fourier transform of $2\pi iyU(y)$.

R9b:

$$U(y) = \int_{-\infty}^{\infty} u(x) \exp(-2\pi ixy)\, dx$$

$$U'(y) = \int_{-\infty}^{\infty} -2\pi ixu(x) \exp(-2\pi ixy)\, dx$$

so $U'(y)$ is the Fourier transform of $-2\pi ixu(x)$.

R10a:

$$\int_{-\infty}^{x} u(\xi)\, d\xi = \int_{-\infty}^{\infty} u(\xi) h(x-\xi)\, d\xi = u(x) \otimes h(x)$$

Taking the (inverse) transform (using R7b) gives

$$U(y)\left[\frac{\delta(y)}{2} + \frac{1}{2\pi iy}\right]$$

where we have used P2a, given below.

R10b:

$$\int_{-\infty}^{y} U(\eta)\, d\eta = \int_{-\infty}^{\infty} U(\eta) h(y-\eta)\, d\eta = U(y) \otimes h(y)$$

Taking the transform gives

$$u(x)\left[\frac{\delta(x)}{2} - \frac{1}{2\pi ix}\right]$$

where we have used P2b, given below.

2B.2 Pairs

P1a:

A derivation, using P5 below (and R5), is given in Section 1.4.

P1b:

This follows from P1a, with R4 and using $\delta(-x) = \delta(x)$.

P2a:

Defining the signum function by

$$\operatorname{sgn}(x) = \begin{cases} 1 & \text{for } x > 0 \\ -1 & \text{for } x < 0 \end{cases} \qquad (x \in \mathbb{R})$$

[and $\operatorname{sgn}(0) = 0$] the unit step function h can be written as

$$2h(x) = 1 + \operatorname{sgn}(x)$$

We now require the transform of sgn which can be given by expressing the signum function as the limit of an antisymmetric decaying exponential function, with form $-\exp(\lambda x)$ for $x < 0$ and $\exp(-\lambda x)$ for $x > 0$ (and $\lambda > 0$):

$$\lim_{\lambda \to 0} \left[\int_{-\infty}^{0} -\exp(\lambda x) \exp(-2\pi ixy)\, dx + \int_{0}^{\infty} \exp(-\lambda x) \exp(-2\pi ixy)\, dx \right]$$

$$= \lim_{\lambda \to 0} \left[- \frac{\exp(\lambda x - 2\pi ixy)}{\lambda - 2\pi iy} \Bigg|_{-\infty}^{0} - \frac{\exp(-\lambda x - 2\pi ixy)}{\lambda + 2\pi iy} \Bigg|_{0}^{\infty} \right]$$

$$= \lim_{\lambda \to 0} \left(- \frac{1}{\lambda - 2\pi iy} - \frac{-1}{\lambda + 2\pi iy} \right)$$

$$= \frac{1}{\pi iy}$$

The Fourier transform of $h(x)$ is now found to be, using P1b,

$$\frac{1}{2}\left[\delta(y) + \frac{1}{\pi iy} \right]$$

P2b:

From P2a and R4, the transform of

$$\frac{1}{2}\left[\delta(x) + \frac{1}{\pi i x}\right]$$

is $h(-y)$; then we use R2, with $\delta(-x) = \delta(x)$.

P3a:

$$\int_{-\infty}^{\infty} \operatorname{rect}(x)\exp(-2\pi ixy)\,dx = \int_{-1/2}^{1/2} \exp(-2\pi ixy)\,dx$$

$$= \frac{\exp(-2\pi ixy)}{-2\pi iy}\bigg|_{-1/2}^{1/2} = \frac{\exp(-\pi iy) - \exp(\pi iy)}{-2\pi iy}$$

$$= \frac{-2i\sin(\pi y)}{-2\pi iy} = \operatorname{sinc}(y)$$

P3b:

From P3a and R4, using $\operatorname{rect}(-y) = \operatorname{rect}(y)$.

P4:

The transform of $\exp(-x)h(x)$ [or $\exp(-x)$ for $x > 0$] is

$$\int_{0}^{\infty} \exp(-x)\exp(-2\pi ixy)\,dx = -\frac{\exp[-(1+2\pi iy)x]}{1+2\pi iy}\bigg|_{0}^{\infty} = +\frac{1}{1+2\pi iy}$$

P5:

$$\int_{-\infty}^{\infty} \exp(-\pi x^2)\exp(-2\pi ixy)\,dx = \int_{-\infty}^{\infty} \exp[-\pi(x+iy)^2 - \pi y^2]\,dx$$

$$= \exp(-\pi y^2)\int_{-\infty+iy}^{\infty+iy} \exp(-\pi z^2)\,dz$$

where $z = x + iy$. We perform a contour integration round the contour shown in Figure 2B.1; as there are no poles within the contour, the contour integral is zero, and as the contributions at $z = \pm\infty + i\eta$ $(0 \leq \eta \leq y)$ are zero, we have

$$\int_{-\infty}^{\infty} \exp(-\pi z^2)\, dz - \int_{-\infty+iy}^{\infty+iy} \exp(-\pi z^2)\, dz = 0$$

so the required integral is equal to the real integral $\int_{-\infty}^{\infty} \exp(-\pi z^2)\, dz$ which has the value 1.

P6–P10:

These are all found using the earlier pairs and rules, as indicated in the Notes column of Table 2.2.

P11:

From the definitions (2.13) and (2.14) we can express the comb function for a constant as a rep function:

$$\text{comb}_X(1) = \sum_{n=-\infty}^{\infty} \delta(x - nX) = \text{rep}_X\, \delta(x)$$

and then, by P1a and R8b, the transform is $|Y|\text{comb}_Y(1)$.

A more rigorous approach is taken in Lighthill [2], particularly for the derivations of the transform of the δ-function, P1a, the transform of the signum function used in obtaining P2a, and the comb and rep transforms.

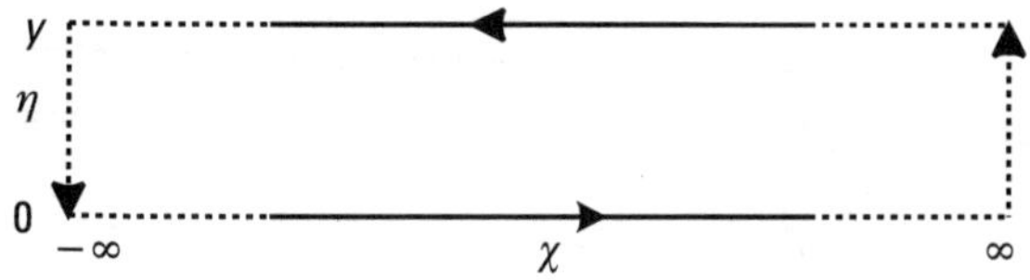

Figure 2B.1 Contour for integral.

3

Pulse Spectra

3.1 Introduction

In this chapter we consider the spectra of pulses and pulse trains. Signals used in radar, sonar, and radio and telephone communications often turn out to be combinations of certain quite simple basic waveforms or of variations on them. For example, the rectangular pulse is an almost universal feature of radar waveforms, and although the perfect pulse is a mathematical idealization, it is often closely realized in practice, and the approximation is good enough for an analysis based on the idealization to give very useful results (which in some cases are obtained very simply).

One reason for studying the spectrum of a pulse, or pulse train, can be to investigate the interference that the pulse transmission will generate outside the frequency band allocated. The sharp-edged rectangular pulse is particularly poor in this respect, producing quite high interference levels at frequencies several times the radar bandwidth away from the radar operating frequency. The interference levels can be lowered quite considerably by reducing the sharp, vertical edges in various ways. Giving the edges a constant finite slope so that the pulse becomes trapezoidal produces a considerable improvement, as shown in Section 3.2. The triangular pulse (Section 3.3) is a limiting case of the trapezoidal with the flat top reduced to zero. The asymmetric trapezoidal pulse (with sides of different magnitude slope) is considered in Section 3.4. While the practical use of such a pulse is not obvious, this is an interesting exercise in the use of the rules-and-pairs method, showing that the method gives a solution for the spectrum quite easily and concisely once a suitable approach has been found. Another form of pulse, smoother than the rectangular pulse, is the raised cosine, and this is shown to have considerably improved spectral side lobes (Section 3.5). The trapezoidal pulse still has sharp corners, and rounding these is the subject of Sections 3.6 and 3.7. Finally, the spectra of pulse trains, as might be used in radar, are studied in the next three sections.

3.2 Symmetrical Trapezoidal Pulse

The rectangular pulse, with zero rise and fall times, may be a reasonable approximation in many cases, but for short pulses the rise and fall times may not be negligible compared with the pulse width and may need to be taken into account. The symmetrical trapezoidal pulse is particularly easily analyzed by the methods used here. We noted in Chapter 2 (Figure 2.7) that such a pulse, of width T between the half amplitude points and with rise and fall times of τ, can be expressed as the convolution of rect functions (illustrated in Figure 3.1):

$$u(t) = (1/\tau)\,\mathrm{rect}\,(t/\tau) \otimes A\,\mathrm{rect}\,(t/T) \tag{3.1}$$

The scaling factor $1/\tau$ keeps the peak height the same, as the narrow pulse now has unit area, though often we are not as interested in the scaling factors as the shapes and relative levels of the waveforms and spectra. The rise and fall times of the edges are τ and the pulse is of width T at the half amplitude points. The spectrum (from R7b, P3a, and R5) is

$$U(f) = AT\,\mathrm{sinc}\,(f\tau)\,\mathrm{sinc}\,(fT) \tag{3.2}$$

Thus the spectrum is that of the pulse of length T multiplied by the broader sinc $f\tau$ function, the transform of the shorter pulse. This will narrow

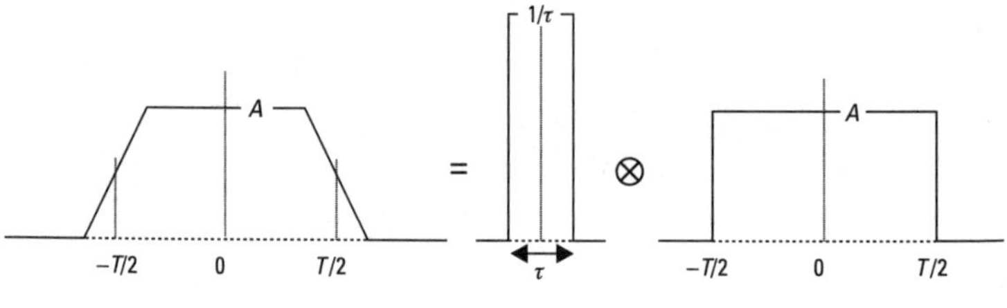

Figure 3.1 Symmetrical trapezoidal pulse.

slightly the width of the main lobe of the spectrum and also reduce the side-lobe levels as shown in Figure 3.2, where $\tau = 0.4T$ in this example.

An application of this result would be to answer (approximately) the question of what rise time, relative to the half-amplitude width, will minimize the first side lobes of the spectrum. We note that the function sinc fT has zeros at $\pm 1/T$ and $\pm 2/T$, and the first side lobes peak at about $\pm 3/2T$. Clearly, we will be very close to minimizing the first side lobes if we make the first zeros of the sinc $f\tau$ function occur at these points. Thus, we require

$$1/\tau = 3/2T, \text{ or } \tau = 2T/3 \tag{3.3}$$

This is not, of course, the precisely optimized solution, but this approximate result is close to optimum and is very easily solved by these methods. In fact, the peak spectral side lobes are 28.8 dB below the peak in this case, compared with only 13.3 dB for the rectangular pulse. If we chose $\tau = 0.6992T$, corresponding to placing the first null of the wide sinc more precisely at the position of the first peak of the narrow sinc (at $\pm 1.4303/T$), then we improve the side lobe discrimination slightly to 30.7 dB. This spectrum is illustrated in Figure 3.3, with the spectrum of the rectangular pulse shown by dotted lines for comparison. (The horizontal axis is in units of $1/T$.)

3.3 Symmetrical Triangular Pulse

A pulse of this shape may arise in practice as a result of convolving rectangular pulses in the process of demodulating a spread spectrum waveform, for example (Figure 3.4). It is the limiting version of the trapezoidal pulse and is given by

$$u(t) = (1/\tau)\,\text{rect}\,(t/T) \otimes A\,\text{rect}\,(t/T) \tag{3.4}$$

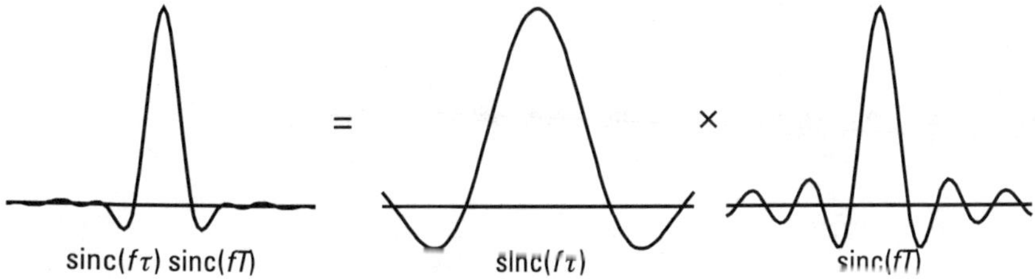

Figure 3.2 Product of sinc functions.

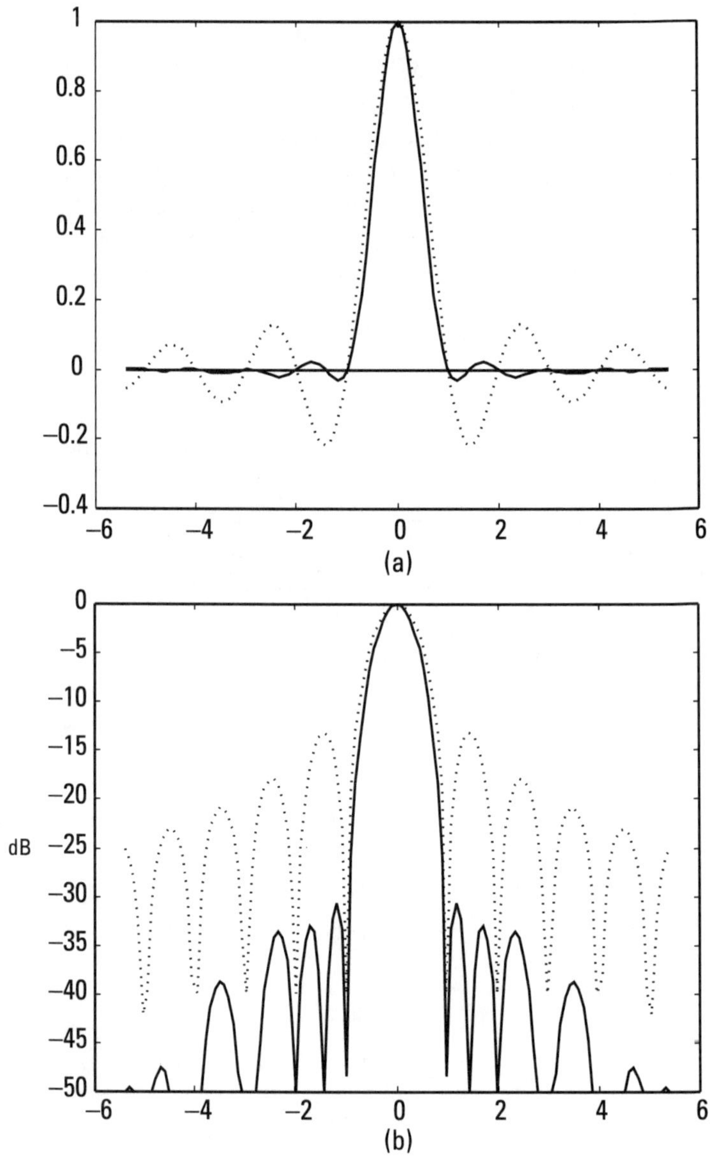

Figure 3.3 Spectrum of low side-lobe trapezoidal pulse: (a) linear form; and (b) logarithmic form.

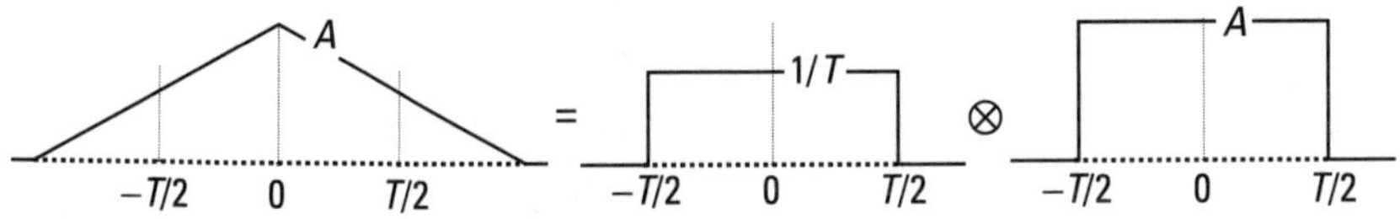

Figure 3.4 Symmetrical triangular pulse.

with spectrum [from (3.2) with $\tau = T$, for example]

$$U(f) = AT\,\mathrm{sinc}^2(fT) \tag{3.5}$$

This is the amplitude spectrum. The power spectrum is a sinc^4 function and is shown in logarithmic form in Figure 3.5, with the rect pulse spectrum for comparison shown dotted. This spectrum has its 3-dB points at $\pm 0.32/T$, its value at $\pm 1/2T$ is nearly 8 dB below the peak value, and the maximum side lobes are 26.5 dB below the peak.

In a case where triangular pulses are used frequently, it could be useful to define a triangular function tri such that

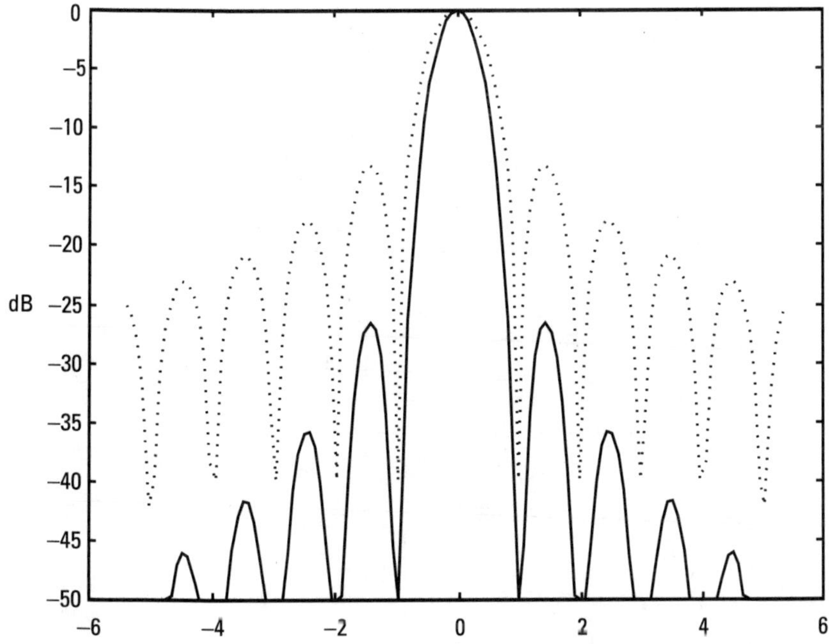

Figure 3.5 Spectrum of triangular pulse.

$$\text{tri}(x) = \begin{cases} 1 + x & \text{for } -1 < x \le 0 \\ 1 - x & \text{for } 0 \le x < 1 \\ 0 & \text{otherwise} \end{cases} \tag{3.6}$$

and then we have

$$\text{tri}(x) = \text{rect}(x) \otimes \text{rect}(x) \tag{3.7}$$

and the transform pair

$$\text{tri}(x) \Leftrightarrow \text{sinc}^2(x) \tag{3.8}$$

3.4 Asymmetrical Trapezoidal Pulse

A linear rising edge of duration τ is given by the convolution of a step function and a pulse of duration τ (Figure 3.6). If the height of the edge is to remain at the same level as the step function, then the convolving pulse must have a height of $1/\tau$. With these results we can define the asymmetric pulse of unit height by the difference of two such modified step functions (Figure 3.7). These have rising edges of the required duration and are centered at $-T/2$ and $T/2$. The pulse is centered, at its half amplitude points, at the origin, and is of width T at this level, with rise and fall times of τ_1 and τ_2. The waveform is given by

$$u(t) = \frac{1}{\tau_1}\text{rect}\left(\frac{t}{\tau_1}\right) \otimes h\left(t + \frac{T}{2}\right) - \frac{1}{\tau_2}\text{rect}\left(\frac{t}{\tau_2}\right) \otimes h\left(t - \frac{T}{2}\right) \tag{3.9}$$

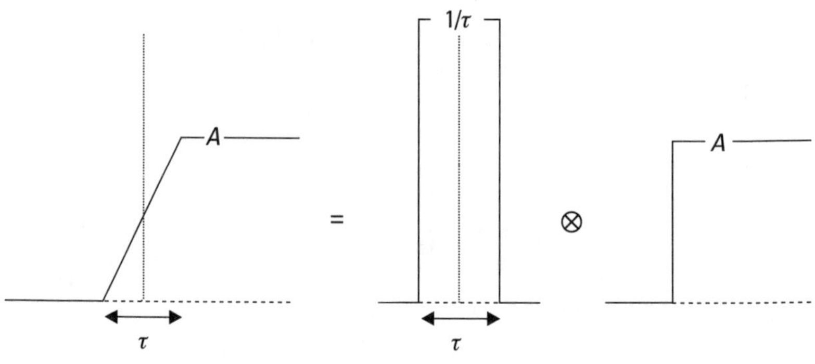

Figure 3.6 Rising edge of width τ.

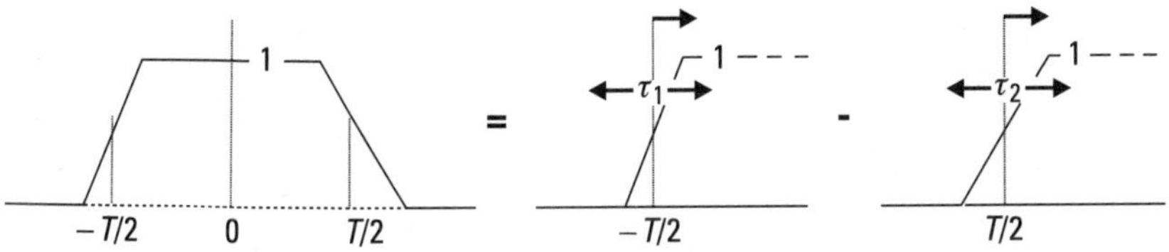

Figure 3.7 Asymmetrical trapezoidal pulse.

The Fourier transform of this waveform is given, using P2a and R6a in addition to the now more familiar P3a and R5, by

$$U(f) = \operatorname{sinc}(f\tau_1)\left[\frac{\delta(f)}{2} + \frac{1}{2\pi i f}\right]e^{\pi i f T} \tag{3.10}$$
$$- \operatorname{sinc}(f\tau_2)\left[\frac{\delta(f)}{2} + \frac{1}{2\pi i f}\right]e^{-\pi i f T}$$

Now, in general, $y(x)\,\delta(x - x_0) = y(x_0)\,\delta(x - x_0)$, as the δ-function is zero except at x_0, so we can put $\operatorname{sinc}(f\tau_1)\,\delta(f)\,e^{\pi f t} = \operatorname{sinc}(0)\,\delta(f)\,e^0 = \delta(f)$ and similarly for the $\operatorname{sinc}(f\tau_2)\,\delta(f)$ term, so that the δ-function terms cancel and we have

$$U(f) = \frac{\operatorname{sinc}(f\tau_1)\,e^{\pi i f T} - \operatorname{sinc}(f\tau_2)\,e^{-\pi i f T}}{2\pi i f} \tag{3.11}$$

We note that the spectrum for the unit height symmetrical pulse, given by putting $\tau_1 = \tau_2 = \tau$ in this expression, is

$$U(f) = \frac{\operatorname{sinc}(f\tau)\left(e^{\pi i f T} - e^{-\pi i f T}\right)}{2\pi i f} = \frac{\operatorname{sinc}(f\tau)\sin(\pi f T)}{\pi f}$$
$$= T\operatorname{sinc}(f\tau)\operatorname{sinc}(fT)$$

which is the result given in Section 3.2 above. Equation (3.11) is a neat and compact expression for the spectrum of this asymmetric function and is very easily found by these methods.

Two examples of the spectrum of an asymmetric pulse are given in Figure 3.8. Only the positive frequency side is given, as these power spectra, of real waveforms, are symmetric about zero frequency, as discussed in Chapter 2 (Section 2.3). The frequency scale is in units of $1/T$, where T is

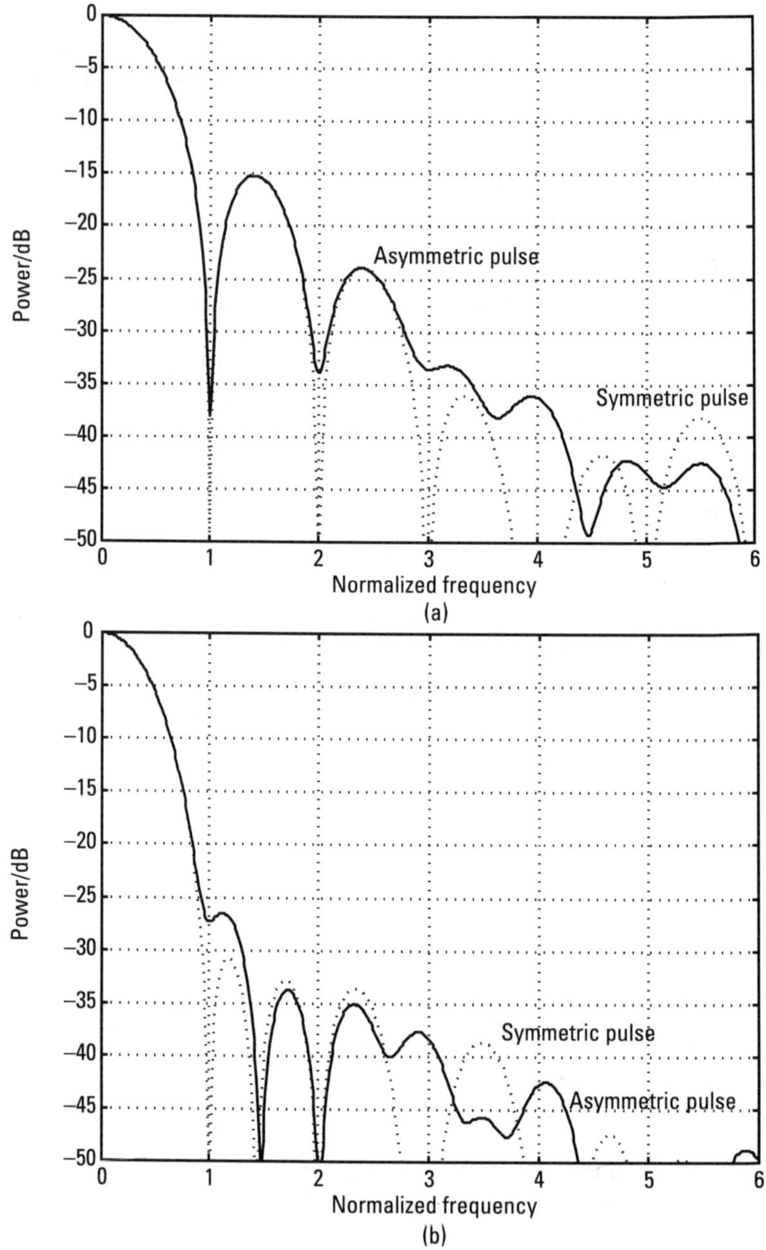

Figure 3.8 Asymmetric trapezoidal pulse spectra: (a) edges 0.2*T* and 0.3*T*; and (b) edges 0.6*T* and 0.8*T*.

the half amplitude pulse width, and, for comparison, the spectra of the symmetric pulses, with rise and fall times equal to the mean of those of the asymmetric pulses, are shown by dotted curves. This mean in the second example, shown in Figure 3.8(b), is $0.7T$, which is very close to the value found in Section 3.2, which places the null due to the slope at the peak of the first side lobe of the underlying rectangular pulse, the spectrum of which is given in Figure 3.3(b). We see that the asymmetry has raised the low first side lobe about 4 dB, while with the sharper edges and higher side lobes of Figure 3.8(a), the effect of asymmetry is not seen until considerably further out in the pattern.

3.5 Raised cosine Pulse

We define this pulse as being of width T at the half amplitude points, which is consistent with the definitions of the triangular and trapezoidal pulses above. Then a unit amplitude pulse is part of the waveform $(1 + \cos 2\pi f_0 t)/2$, where $f_0 = 1/2T$ (i.e., $2T$ is the duration of a cycle of the cosine). This waveform is gated for a time $2T$, so the pulse is given by

$$u(t) = \text{rect}\,(t/2T)\,(1 + \cos\, 2\pi f_0 t)/2 \qquad (3.12)$$

The unit amplitude pulse is shown in Figure 3.9(a), with the time axis in units of T.

The spectrum is thus, using P3a, P1b, P7a, and R5,

$$\begin{aligned} U(f) &= 2T\,\text{sinc}\,2fT \otimes \frac{\delta(f) + \frac{1}{2}\,[\delta(f - f_0) + \delta(f + f_0)]}{2} \\ &= T\left(\text{sinc}\,(f/f_0) + \frac{1}{2}\,\text{sinc}\,[(f - f_0)/f_0] + \frac{1}{2}\,\text{sinc}\,[(f + f_0)/f_0]\right) \end{aligned} \qquad (3.13)$$

as convolution with a δ-function corresponds to a shift in the position of the sinc function.

This spectrum is not narrowband, as the "carrier" frequency f_0 is not large compared with the "modulation" bandwidth and the spectrum is seen to consist of three closely overlapping sinc functions. These are shown as dotted lines in Figure 3.9(b), with the pulse spectrum as the solid line. The

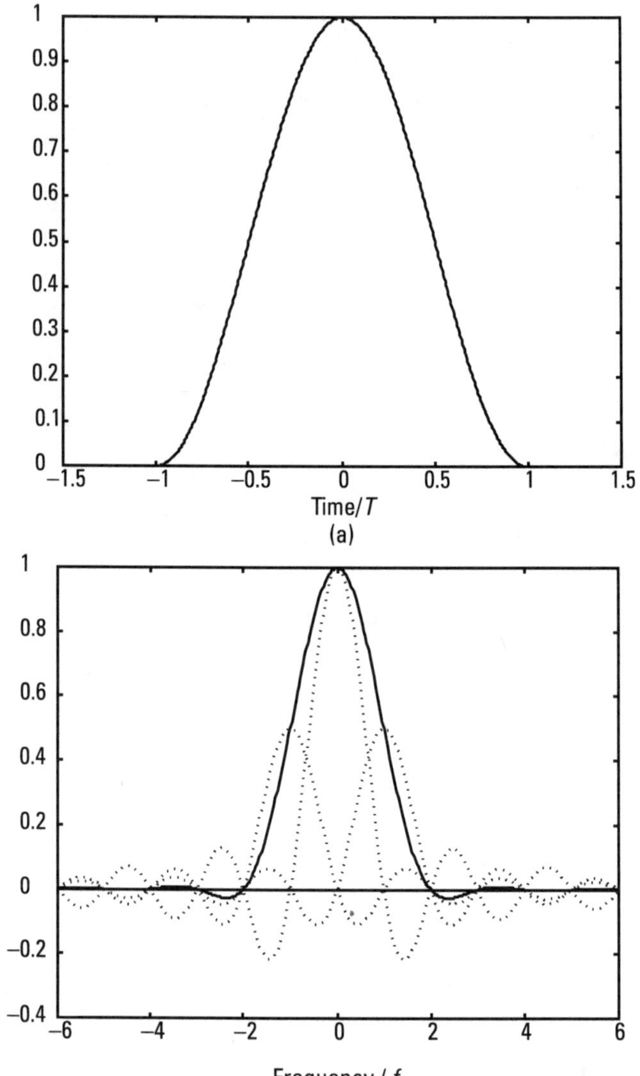

Figure 3.9 Raised cosine pulse: (a) normalized waveform; and (b) normalized spectrum.

frequency axis is in units of f_0 or $1/2T$. These sum to give a spectral shape with first zeros at $\pm 2f_0$ or $\pm 1/T$ (and zeros in general at $n/2T$ for n integral, $|n| \geq 2$) with quite low spectral side lobes. These are shown more clearly in logarithmic form in Figure 3.10, with the spectrum of the gating pulse for comparison. The highest spectral side lobes are 31 dB below the peak.

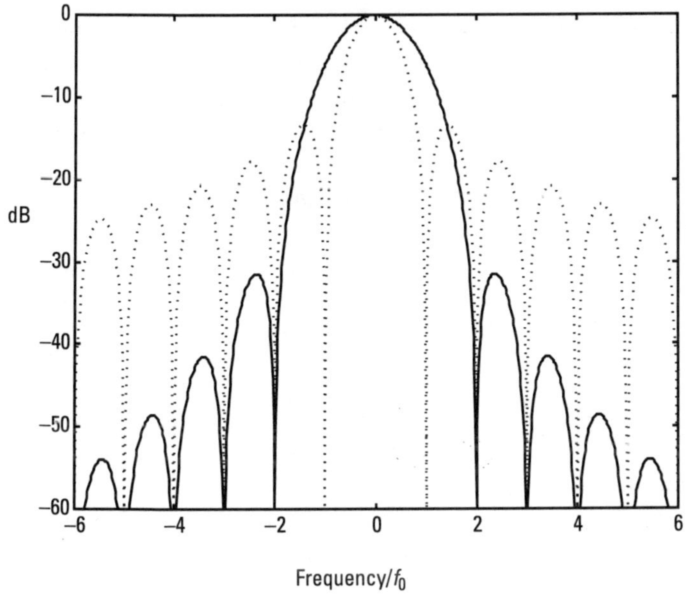

Figure 3.10 Raised cosine pulse spectrum, log scale.

These lower side lobes could be expected from the much smoother shape of this pulse, compared with the rectangular or triangular pulses, the highest side lobes of which are 13 dB and 27 dB below the peak, respectively. We note that the cost of lower side lobes is a broadening of the main lobe relative to the spectrum of the gating pulse of width $2T$. The broadening is by a factor of 1.65 at the 4-dB points.

3.6 Rounded Pulses

The step discontinuity of rectangular pulses is the cause of the poor spectrum, with high side lobes. This discontinuity in level is removed by generating rising and falling edges of finite slope. In the case of the symmetric trapezoidal pulse, this is achieved by the convolution of the rectangular pulse with another, shorter rectangular pulse, as shown in Section 3.2. This reduction in discontinuity improves the side-lobe levels. There are still discontinuities in slope for these pulses, and these can be removed by another convolution, with a further reduction in side-lobe levels. The convolution need not, in principle, be with a rectangular pulse, but this is perhaps the simplest and is the example taken here.

Figure 3.11 illustrates the effect of convolution with a rectangular pulse on one of the corners of the trapezoidal pulse. The pulse is of length T and over the region $-T/2$ to $+T/2$ relative to the position of the corner the waveform rises as t^2, returning to a constant slope (rising as t) after this interval.

Convolving the trapezoidal pulse with this rectangular pulse will round all four corners in a similar manner. If $f(t)$ describes the trapezoidal pulse waveform and $F(f)$ is its spectrum, then for the rounded waveform we have (from R7b, P3a, and R5)

$$f(t) \otimes (1/T)\,\mathrm{rect}\,(t/T) \Leftrightarrow F(f)\,\mathrm{sinc}\,fT \tag{3.14}$$

that is, the spectrum is multiplied by the spectrum of the short pulse, which lowers the side lobes further.

In practice, a pulse is likely to be rounded by stray capacity, which could be modeled by the circuit shown in Figure 3.12.

In electrical engineering notation, the frequency response of this network is given by

$$A(\omega) = \frac{(1/R_2 + j\omega C)^{-1}}{R_1 + (1/R_2 + j\omega C)^{-1}} = \frac{1}{1 + R_1/R_2 + j\omega CR_1} \tag{3.15}$$

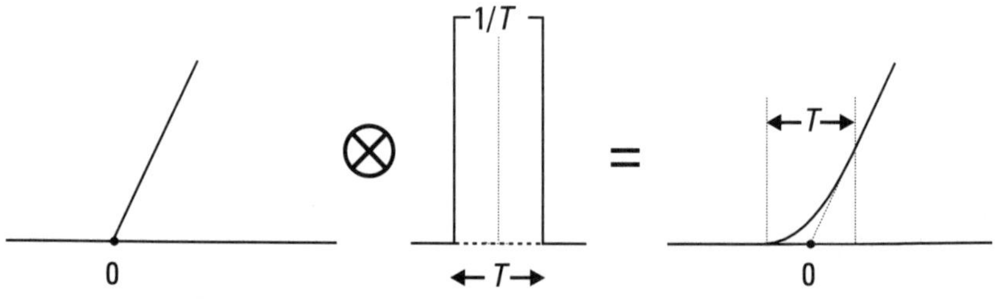

Figure 3.11 Rounded corner of width T.

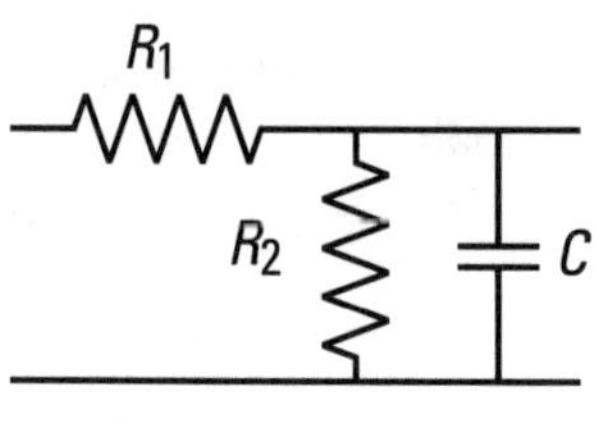

Figure 3.12 Filter model.

where $j^2 = -1$ and ω is the angular frequency $2\pi f$. In the notation we use here, this becomes

$$A(f) = \frac{1}{1 + R_1/R_2 + 2\pi ifCR_1} = \frac{R_2}{R_1 + R_2} \cdot \frac{1}{1 + 2\pi if\tau} \quad (3.16)$$

where

$$\tau = \frac{CR_1R_2}{R_1 + R_2} \quad (3.17)$$

The product of capacitance and resistance has the dimension of time, so τ represents a time constant for the circuit, and the factor $R_2/(R_1 + R_2)$ is the limiting attenuation to low-frequency signals (approaching dc or $f = 0$).

The impulse response $a(t)$ of this circuit is the (inverse) Fourier transform of the frequency response, and from P4 and R5, we have [apart from the scaling factor $R_2/(R_1 + R_2)$]

$$a(t) = \frac{1}{\tau} e^{-t/\tau} \ (t \geq 0) \text{ or } a(t) = \frac{1}{\tau} e^{-t/\tau} h(t) \quad (3.18)$$

where h is the step function. To investigate the form of the rounding this produces, we form the convolution of $a(t)$ with the rising edge, given by $e(t) = kt \cdot h(t)$. This is

$$a(t) \otimes e(t) = \int_{-\infty}^{\infty} \frac{1}{\tau} e^{-(t-t')/\tau} h(t - t') kt'h(t') \, dt' = \int_{0}^{t} \frac{k}{\tau} t' e^{-(t-t')/\tau} \, dt'$$
$$= k[t - \tau(1 - e^{-t/\tau})] \quad (3.19)$$

For large t, the exponential term becomes small, and we see that the response approaches $k(t - \tau)$, corresponding to a delay of τ applying to the pulse as whole (in addition to the rounding distortion), assuming τ is small compared with the pulse duration. We note that if we move the rectangular rounding pulse of Figure 3.11 so that it starts, like the exponential impulse response, at time zero rather than at $-T/2$, then this rectangular "impulse response" causes a delay of $T/2$, so this pulse with length $T = 2\tau$ will give the same delay and will be approximately equivalent. Figure 3.13 shows the

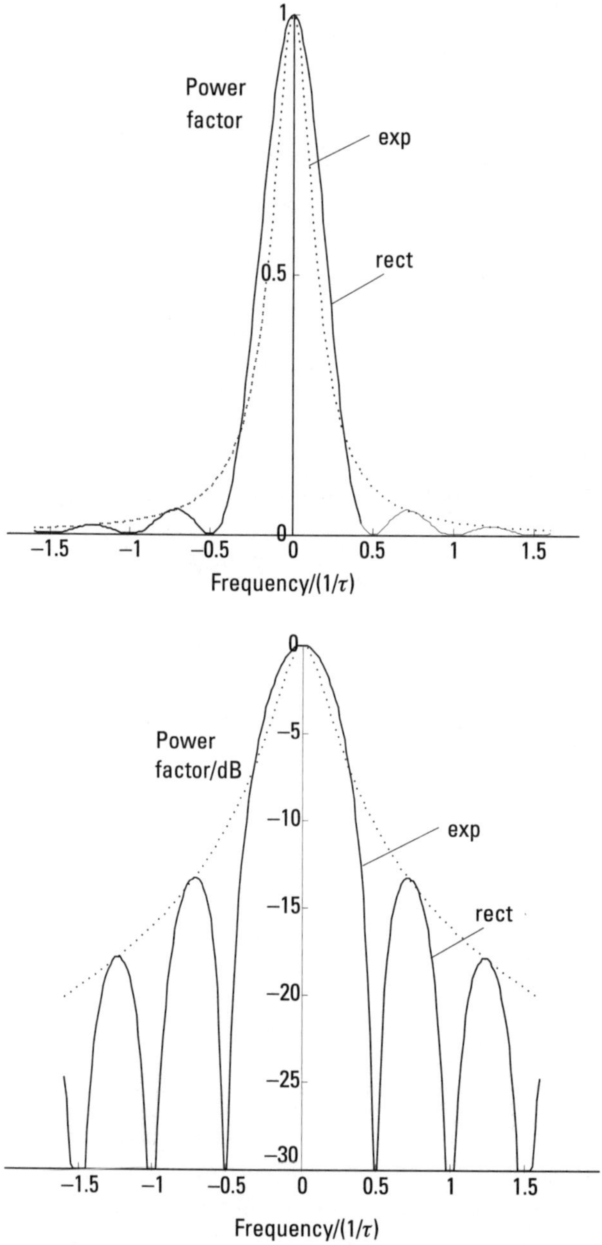

Figure 3.13 Power spectra for rect and exponential impulse responses.

spectral power factors (in both linear and logarithmic form) multiplying the original pulse spectrum in the two cases, $\text{sinc}^2 2f\tau$ for the rectangular pulse and $1/[1 + (2\pi f\tau)^2]$ for the stray capacitance. The power spectrum of the smoothed pulse is that of the spectrum of the original pulse multiplied by one of these spectra. Assuming the smoothing impulse response is fairly short compared with the pulse length, the spectrum of the pulse will be mainly within the main lobe of the impulse response spectrum. We see that the side-lobe pattern of the pulse will be considerably reduced by the smoothing (e.g., by about 10 dB at $\pm 0.4/\tau$ from center frequency). We also see that the rect pulse of width 2τ gives a response fairly close to the stray capacitance filter with time constant τ.

3.7 General Rounded Trapezoidal Pulse

Here we consider the problem of rounding the four corners of a trapezoidal pulse over different time intervals. This may not be a particularly likely problem to arise in practice in connection with radar, but the solution to this awkward case is interesting and illuminating, and could be of use in some other application.

The problem of the asymmetrical trapezoidal pulse was solved in Section 3.4 by forming the pulse from the difference of two step-functions, each of which was convolved with a rectangular pulse to form a rising edge. By using different-width rectangular pulses, we were able to obtain different slopes for the front and back edges of the pulse.

In this case we extend this principle by expressing the convolving rect pulses themselves as the difference of two step functions. The (finite) rising edge can then be seen to be the difference of two infinite rising edges, as shown in Figure 3.14. Each of these, which we call Ramp functions, is produced by the convolution of two unit step functions as shown in Figure 3.15 and defined in (3.20) below.

We define the Ramp function, illustrated in Figure 3.15, by

$$\text{Ramp}(t - T) = h(t) \otimes h(t - T) \tag{3.20}$$

so that

$$\text{Ramp}(t) = \begin{cases} 0 & \text{for } t \leq 0 \\ t & \text{for } t > 0 \end{cases} \qquad (t \in \mathbb{R}) \tag{3.21}$$

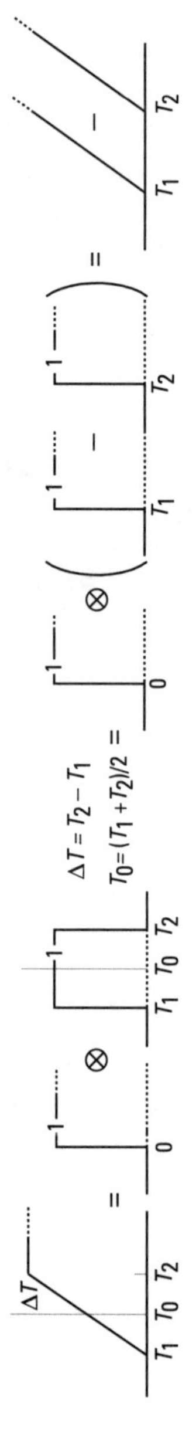

Figure 3.14 Rising edge as the difference of two Ramp functions.

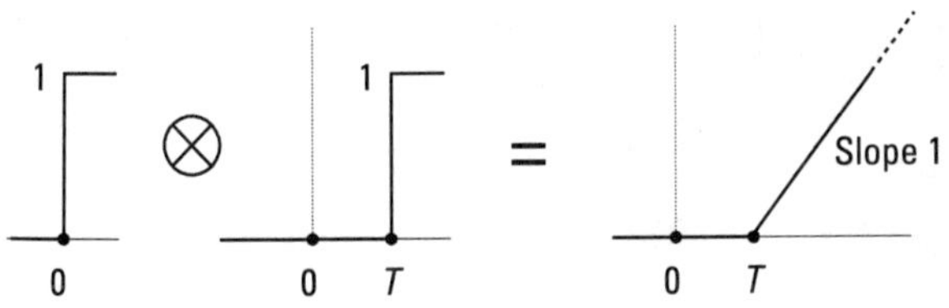

Figure 3.15 Ramp function.

(A different, finite, linear function is required in Chapter 6; this is called ramp.) Having now separated the four corners of the trapezoidal pulse into the corners of four Ramp functions, they can now all be rounded separately by convolving the Ramp functions with different-width rect functions (or other rounding functions, if required) as in Figure 3.11, before combining to form the smoothed pulse. Before obtaining the Fourier transform of the rounded pulse, we obtain the transform of the trapezoidal pulse in the form of the four Ramp functions (two for each of the rising and falling edges).

In mathematical notation, the rising edge of Figure 3.14 can be expressed in the two ways

$$h(t) \otimes \operatorname{rect}\left(\frac{t - T_0}{\Delta T}\right) = h(t) \otimes (\operatorname{Ramp}(t - T_1) - \operatorname{Ramp}(t - T_2)) \tag{3.22}$$

The Fourier transform of the left side is, from P2a, P3a, R7b, R5, and R6a,

$$\left[\frac{\delta(f)}{2} + \frac{1}{2\pi i f}\right] \Delta T \operatorname{sinc} f\Delta T \exp(-2\pi i f T_0) \tag{3.23}$$

$$= \Delta T\left[\frac{\delta(f)}{2} + \frac{\operatorname{sinc} f\Delta T \exp(-2\pi i f T_0)}{2\pi i f}\right]$$

where we have used $\delta(f - f_0)u(f) = \delta(f)u(f_0)$ in general, so $\delta(f)\operatorname{sinc}(f\Delta T) = \delta(f)$. The transform of the difference of the Ramp functions on the right side is, using (3.20), P2a, R7b, and R6a,

$$\left[\frac{\delta(f)}{2} + \frac{1}{2\pi i f}\right]\left\{\left[\frac{\delta(f)}{2} + \frac{1}{2\pi i f}\right][\exp(-2\pi i f T_1) - \exp(-2\pi i f T_2)]\right\} \tag{3.24}$$

Using T_0 and ΔT as given in Figure 3.14, the difference of the exponential terms becomes $\exp(-2\pi i f T_0)(\exp(2\pi i f \Delta T) - \exp(-2\pi i f \Delta T))$ or $2i \sin(2\pi f \Delta T)\exp(-2\pi i f T_0)$, so again using $\delta(f - f_0)u(f) = \delta(f)u(f_0)$ [with $u(f_0) = \sin(0)$ in this case], (3.24) becomes

$$\left[\frac{\delta(f)}{2} + \frac{1}{2\pi i f}\right]\left\{\left[\frac{\delta(f)}{2} + \frac{1}{2\pi i f}\right] 2i \sin(\pi f \Delta T)\exp(-2\pi i f T_0)\right\}$$

$$= \left[\frac{\delta(f)}{2} + \frac{1}{2\pi i f}\right]\frac{\sin(\pi f \Delta T)\exp(-2\pi i f T_0)}{\pi f} \tag{3.25}$$

$$= \left[\frac{\delta(f)}{2} + \frac{1}{2\pi i f}\right]\Delta T \operatorname{sinc}(f \Delta T)\exp(-2\pi i f T_0)$$

which is the same as (3.23), as expected.

We are now in a position to find the spectrum of the trapezoidal pulse shown in Figure 3.16, with different roundings of each corner. This pulse is separated, as shown, into four Ramp functions and has rising and falling edges of width ΔT_r and ΔT_f, centered at T_r and T_f, respectively. The edges, formed from pairs of Ramp functions, are normalized to unity by dividing by the width ΔT_r or ΔT_f. (They certainly have to be scaled to the same height if the initial and final levels are to be the same.) Thus this pulse is given by

$$\frac{1}{\Delta T_r}[\mathrm{Ramp}(t - T_1) - \mathrm{Ramp}(t - T_2)] \tag{3.26}$$
$$- \frac{1}{\Delta T_f}[\mathrm{Ramp}(t - T_3) - \mathrm{Ramp}(t - T_4)]$$

To round a corner we replace $\mathrm{Ramp}(t - T_k)$ by $r_k(t) \otimes \mathrm{Ramp}(t - T_k)$, where $r_k(t)$ is a rounding function of unit integral (such as the rect pulse in Figure 3.11). For a function with this property, it follows that $R(0) = 1$, where R is the Fourier transform of r; this is shown by

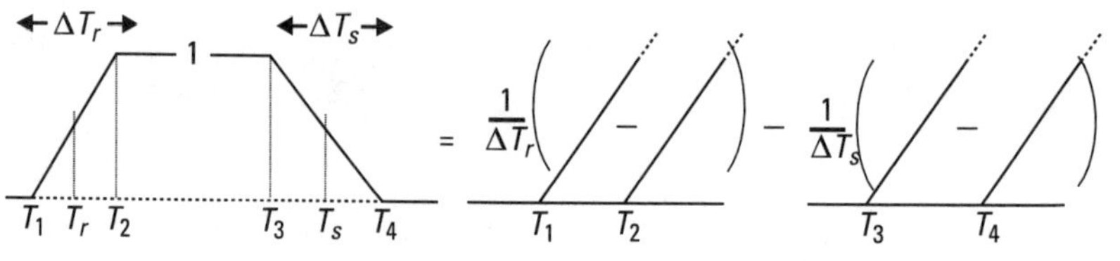

Figure 3.16 Unit height trapezoidal pulse.

$$\int_{-\infty}^{\infty} r(t)\,dt = 1 = \int_{-\infty}^{\infty} r(t)\,e^{-2\pi ift}\,dt\bigg|_{f=0} = R(0) \tag{3.27}$$

The rounded rising edge, given by $e_r(t) = [r_1(t) \otimes \text{Ramp}(t - T_1) - r_2(t) \otimes \text{Ramp}(t - T_2)]/\Delta T_r$, can be written, from the definition of Ramp in (3.20),

$$e_r(t) = h(t) \otimes [r_1(t) \otimes h(t - T_1) - r_2(t) \otimes h(t - T_2)]/\Delta T_r \tag{3.28}$$

with transform

$$\begin{aligned}
E_r(f) &= \frac{1}{\Delta T_r}\left[\frac{\delta(f)}{2} + \frac{1}{2\pi if}\right] \\
&\quad \cdot \left\{\left[\frac{\delta(f)}{2} + \frac{1}{2\pi if}\right][R_1(f)\exp(-2\pi ifT_1) - R_2(f)\exp(-2\pi ifT_2)]\right\} \\
&= \left[\frac{\delta(f)}{2} + \frac{1}{2\pi if}\right] \\
&\quad \cdot \left\{\frac{[R_1(f)\exp(\pi if\Delta T_r) - R_2(f)\exp(-\pi if\Delta T_r)]}{2\pi if\Delta T_r}\exp(-2\pi ifT_r)\right\} \\
&= \left\{\frac{\delta(f)}{2} + \frac{[R_1(f)\exp(\pi if\Delta T_r) - R_2(f)\exp(-\pi if\Delta T_r)]}{(2\pi if)^2\Delta T_r}\right\}\exp(-2\pi ifT_r)
\end{aligned} \tag{3.29}$$

following the approach of the nonrounded case above [(3.23) to (3.25)]. Combining the two edges, the δ-functions disappear, as in forming the spectrum of the asymmetric pulse in Section 3.4 [(3.10) and (3.11)], to give the final result for the spectrum of the generally rounded trapezoidal pulse:

$$\begin{aligned}
&-\frac{\left[R_1(f)e^{\pi if\Delta T_r} - R_2(f)e^{-\pi if\Delta T_r}\right]}{(2\pi f)^2\Delta T_r}e^{-2\pi ifT_r} \\
&+\frac{\left[R_3(f)e^{\pi if\Delta T_s} - R_4(f)e^{-\pi if\Delta T_s}\right]}{(2\pi f)^2\Delta T_s}e^{-2\pi ifT_s}
\end{aligned} \tag{3.30}$$

As a check, we note that if we used a single rounding function r, with transform R, the expression in (3.30) reduces to

$$R(f)\left(\frac{\operatorname{sinc} f\Delta T_r}{2\pi i f} e^{-2\pi i f T_r} - \frac{\operatorname{sinc} f\Delta T_s}{2\pi i f} e^{-2\pi i f T_s}\right) \tag{3.31}$$

which (with $T_r = -T/2$, $T_s = T/2$, $\Delta T_r = \tau_1$, and $\Delta T_s = \tau_2$) is seen, from (3.11), to be exactly the result of smoothing the asymmetrical trapezoidal pulse with the function r.

3.8 Regular Train of Identical RF Pulses

This waveform could represent, for example, an approximation to the output of a radar transmitter using a magnetron triggered at regular intervals. The waveform is defined by

$$u(t) = \operatorname{rep}_T[\operatorname{rect}(t/\tau)\cos 2\pi f_0 t] \tag{3.32}$$

where the pulses of length τ of a carrier at frequency f_0 are repeated at the pulse repetition interval T and shown in Figure 3.17.

We note that the rep operator applies to a product of two functions, so the transform will be (by R8b) a comb version of a convolution of the transforms of these functions. We could express the cosine as a sum of exponentials, but more conveniently we use P7a in which this has already been done. Thus (from P3a, P8a, R8b, and R5) we obtain

$$U(f) = (\tau/2T)\operatorname{comb}_{1/T}[\operatorname{sinc}(f-f_0)\tau + \operatorname{sinc}(f+f_0)\tau] \tag{3.33}$$

This spectrum is illustrated (in the positive frequency region) in Figure 3.18.

Thus we see that the spectrum consists of lines (which follows from the repetitive nature of the waveform) at intervals $1/T$, with strengths given

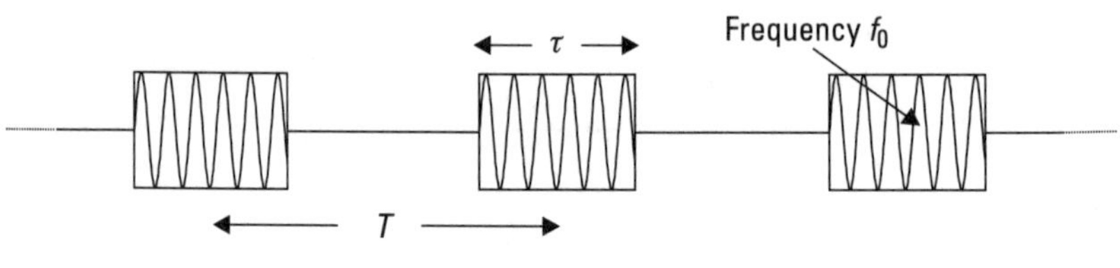

Figure 3.17 Regular train of identical RF pulses.

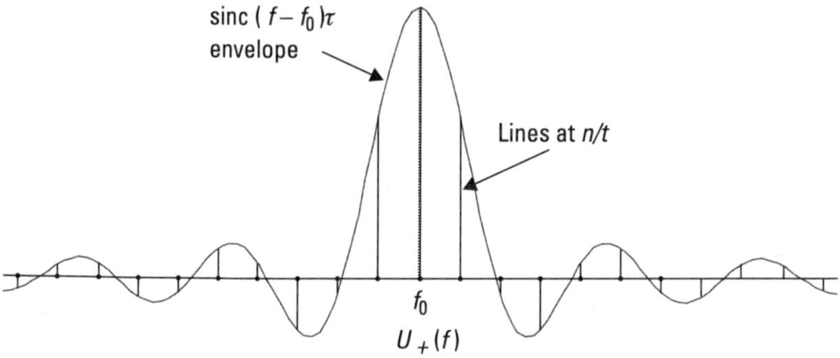

Figure 3.18 Spectrum of regular RF pulse train.

by two sinc function envelopes centered at frequencies f_0 and $-f_0$. As discussed in Chapter 2, the negative frequency part of the spectrum is just the complex conjugate of the real part (for a real waveform) and provides no extra information. (In this case the spectrum is real, so the negative frequency part is just a mirror image of the real part.) However, as explained in Section 2.4.1, the contribution of the part of the spectrum centered at $-f_0$ in the positive frequency region can only be ignored if the waveform is sufficiently narrowband (i.e., if $f_0 >> 1/\tau$, the approximate bandwidth of the two spectral branches).

An important point about this spectrum, which is very easily made evident by this analysis, is that, although the envelope of the spectrum is centered at f_0, there is, in general, no spectral line at f_0. This is because the lines are at multiples of the pulse repetition frequency (PRF) $(1/T)$, and only if f_0 is an exact multiple of the PRF will there be a line at f_0. Returning to the time domain, we would not really expect power at f_0 unless the carrier of one pulse were exactly in phase with the carrier of the next pulse. For there to be power at f_0, there should be a precisely integral number of wavelengths of the carrier in the repetition interval T; that is, the carrier frequency should be an exact multiple of the PRF. This is the case in the next example.

3.9 Carrier Gated by a Regular Pulse Train

This waveform would be used, for example, by a pulse Doppler radar. A continuous stable frequency source is gated to produce the required pulse train (Figure 3.19). Again we take T for the pulse repetition interval, τ for

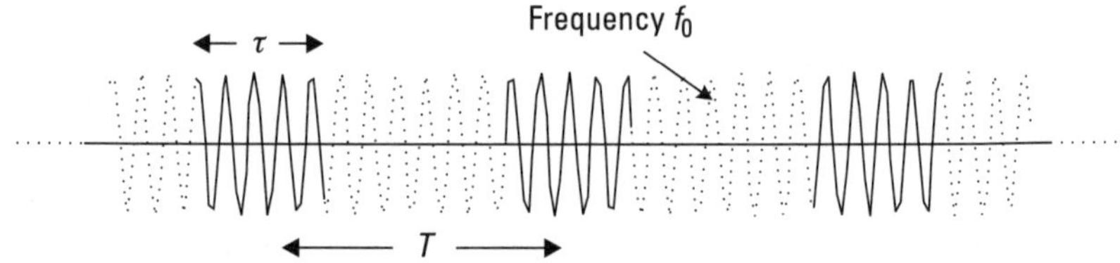

Figure 3.19 Carrier gated by a regular pulse train.

the pulse length, and f_0 for the carrier frequency. The waveform is given by

$$u(t) = [\text{rep}_T(\text{rect } t/\tau)] \cos 2\pi f_0 t \tag{3.34}$$

and its transform, shown in Figure 3.20, is (using R7a, R8b, P3a, and P7a)

$$U(f) = (\tau/2T)\,\text{comb}_{1/T}(\text{sinc } f\tau) \otimes [\delta(f - f_0) + \delta(f + f_0)] \tag{3.35}$$

Denoting the positive frequency part of the spectrum by U_+ and assuming the waveform is narrowband enough to give negligible overlap of the two parts of the spectrum, we have

$$U_+(f) = (\tau/2T)\,\text{comb}_{1/T}(\text{sinc } f\tau) \otimes \delta(f - f_0) \tag{3.36}$$

The function $\text{comb}_{1/T}\text{sinc } f\tau$ is centered at zero and has lines at multiples of $1/T$, including zero. Convolution with $\delta(f - f_0)$ simply moves the center of this whole spectrum up to f_0. Thus there are lines at $f_0 + n/T$

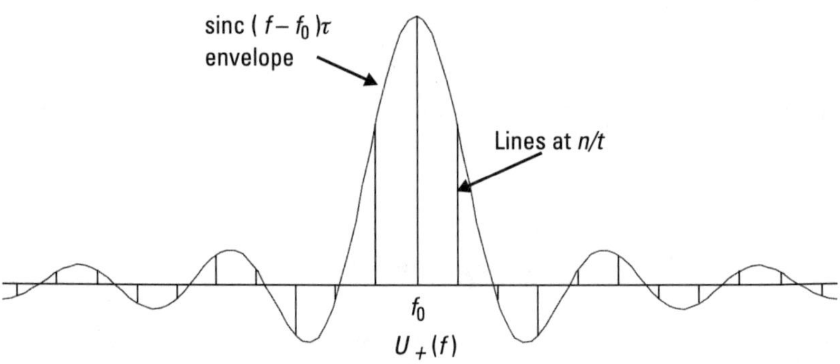

Figure 3.20 Spectrum of regularly gated carrier.

(n integral, $-\infty$ to ∞), including one at f_0. In general, there is no line at $f = 0$; this is only the case if f_0 is an exact multiple of $1/T$. Unlike the previous case, we would expect the waveform to have power at f_0, as the pulses all consist of samples of the same continuous carrier at this frequency.

3.10 Pulse Doppler Radar Target Return

In this case we take the radar model to be a number of pulses with their amplitudes modulated by the beam shape of the radar as it sweeps past the target. Here, for simplicity, we approximate this modulation first by a rectangular function of width θ (i.e., θ is the time on target). A more realistic model will be taken later. The transmitted waveform (and hence the received waveform, from a stationary point target) is given, apart from an amplitude scaling factor, by

$$x(t) = \mathrm{rect}\,(t/\theta)\,u(t) \tag{3.37}$$

where $u(t)$ is given in (3.34) above. The spectrum (from R7a, P3a, and R5) is

$$X(t) = \theta\ \mathrm{sinc}\, f\theta \otimes U(f) \tag{3.38}$$

where U is given in (3.35). The convolution effectively replaces each δ-function in the spectrum U by a sinc function. This is of width $1/\theta$ (at the 4-dB points), which is small compared with the envelope sinc function of the spectrum, which has width $1/\tau$, and also is small compared with the line spacing $1/T$ if $\theta >> T$ (i.e., many pulses are transmitted in the time on target). In fact there will also be a Doppler shift on the echoes if the target is moving relative to the radar. If it has a relative approaching radial velocity v, then the frequencies in the received waveform should be scaled by the factor $(c + v)/(c - v)$, where c is the speed of light. This gives an approximate overall spectral shift of $+2vf_0/c$ (assuming $v << c$ and the spectrum is narrowband, so that all significant spectral energy is close to f_0 or $-f_0$). Figure 3.21 illustrates the form of the spectrum of the received signal. Stationary objects (or "clutter") produce echoes at frequency f_0 and at intervals n/T about f_0, all within an envelope defined by the pulse spectrum (as in Figure 3.20). The smaller, moving target echoes produce lines offset from the clutter lines, so that such targets can be seen, as a consequence of their relative movement, in the presence of otherwise overwhelming clutter.

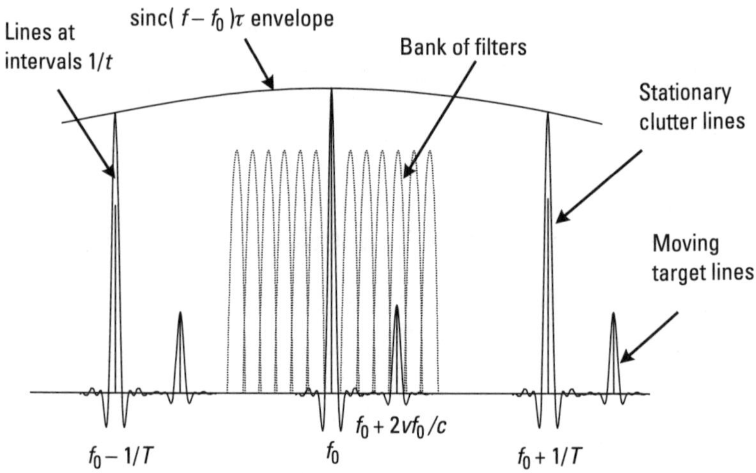

Figure 3.21 Spectrum of pulse Doppler radar waveform.

(Figure 3.21 is diagrammatic; the filter bank may be at baseband or a low IF, and may be realized digitally. By suitable filtering, not only can the targets be seen, but an estimate is obtained of the Doppler shift and hence of the target radial velocity.)

As indicated by (3.38), all the lines are broadened by the spectrum of the beam modulation response. In Chapter 7 we will see that, for a linear aperture, the beam shape is essentially the inverse Fourier transform of the aperture illumination function, and with a constant angular rotation rate, this becomes the beam modulation. (We require the small angle approximation $\sin \alpha \approx \alpha$, which is generally applicable in the radar case.) The transform of this will give essentially the same function as the illumination function. Thus, if this is chosen to be, for example, the raised cosine function (as in Section 3.5) to give moderately low side lobes (Figure 3.10), then the lines will be spread by a raised cosine function, also.

3.11 Summary

The spectra of a number of pulses and of pulse trains have been obtained in this chapter using the rules-and-pairs method. As remarked earlier, the aim is not so much to provide a set of solutions on this topic as to illustrate the use of the method so that users can become familiar with it and then solve their own problems using it. Thus, whether all the examples correspond demonstrably to real problems (for example, finding the spectra of the asymmetric trapezoidal pulse and, particularly, this pulse with different roundings of each corner) is not the question—the variety of possible user problems cannot be anticipated, after all—but rather whether the examples demonstrate various ways of applying the method to yield solutions neatly and concisely without any explicit integration.

4

Sampling Theory

4.1 Introduction

In this chapter we use the rules-and-pairs notation and technique to derive several sampling theorem results, which can be done very concisely in some cases. In fact, the wideband (or baseband) sampling theorem and the Hilbert sampling theorem for narrowband (or RF and IF) waveforms are obtained here following the derivations of Woodward [1]. Two other narrowband sampling techniques, uniform sampling and quadrature sampling, have been analyzed by Brown [2], but these results have been obtained here much more easily using Woodward's approach and have been extended to show what sampling rates are acceptable, rather than just giving the minimum sampling rates presented by Brown.

Woodward's technique is to express the spectrum U of the given waveform u in a repetitive form, then gate it to obtain the spectrum again. The Fourier transform of the resulting identity shows that the waveform can be expressed as a set of impulses of strength equal to samples of the waveform, suitably interpolated. This is the converse of repeating a waveform to obtain a line spectrum: if a waveform is repeated at intervals T, a spectrum is obtained consisting of lines (δ-functions in the frequency domain) at intervals $F = 1/T$ with envelope U, the spectrum of u. Conversely, if a spectrum U is repeated at intervals F, we obtain a waveform of impulses (δ-functions in the time domain) at intervals $T = 1/F$ with envelope u, the (inverse) transform of U. The problem in this case is to express the spectrum precisely as a gated repetitive form of itself. In general, this can only be done by specifying that U should have no power outside a certain frequency interval, and that there should be no overlapping when U is repeated. (In one case below, that of quadrature sampling, overlapping is allowed, provided a condition is met, but again this is for the case of a strictly band-limited spectrum.) This finite bandwidth condition is not a completely realizable one—it corresponds to an infinite waveform—but can be interpreted as the condition that U should have *negligible* power (rather than *no* power) outside the given band. The values that are "negligible" will depend on the system and are not analyzed here. However, the approach used here can be used to determine, or at least to estimate, the effect of spectral overlap, which is, in fact, aliasing.

Brown's approach is to express the waveform u as an expansion in terms of orthogonal time functions. In fact, these orthogonal functions are just the set of displaced interpolating functions of the Woodward approach, the interpolating function being the Fourier transform of the spectral gating function. It is necessary to show that this set of functions, which varies with the sampling technique used, is complete. This method is rather complicated compared with Woodward's, which can use the standard results for Fourier series using sets of complex exponential, or trigonometrical, functions. Furthermore, the Woodward approach seems generally easier to understand and so to modify or apply to other possible sampling methods.

4.2 Basic Technique

First we present the basic technique that is used in subsequent sections to derive the sampling theory results. Because a regularly sampled waveform, which is the ultimate target, has a repetitive spectrum, we repeat the spectrum U of the given waveform u at frequency intervals F, then gate (or filter) this spectrum to obtain U again. This identity is then Fourier transformed to produce an identity between the waveform and an interpolated sampled form of itself. Because this is an identity, it means that all the information in the original waveform u is contained in the sampled form. (The definition of the interpolating function is also needed if it is required to reconstitute the analogue waveform u.) In symbols we write

$$U(f) = \mathrm{rep}_F\, U(f)\, G(f) \tag{4.1}$$

$$u(t) = (1/F)\, \mathrm{comb}_{1/F}\, u(t) \otimes g(t) \tag{4.2}$$

where $G(f)$ is the spectral gating function and $g(t)$ is its transform (i.e., the impulse response of a filter with frequency response G).

Now the comb function consists of a set of impulse responses (δ-functions) at intervals $T = 1/F$ of strength equal to the value of the function u at the instant of the impulse:

$$\text{comb}_T u(t) = \Sigma u(nT)\,\delta(t - nT) \tag{4.3}$$

The convolution of a function g with a δ-function simply transfers the origin of g to the position of the δ-function. Thus (4.2) and (4.3) give, with $T = 1/F$,

$$\begin{aligned} u(t) &= T\,\text{comb}_T u(t) \otimes g(t) = T\Sigma u(nT)\,\delta(t - nT) \otimes g(t) \\ &= T\Sigma u(nT) g(t - nT) \end{aligned} \tag{4.4}$$

This makes clear the identity between u and its sampled form, correctly interpolated.

In the following sections of this chapter, the starting point is (4.1), choosing the appropriate sampling frequency F and spectral gating function G in the different cases. The basic problem is to express U in terms of a gated repetitive form of itself, where the repetition frequency F is chosen so that no spectral overlapping occurs. We are primarily concerned with determining F, which is the required sampling rate, and are less concerned with the gating function G, except that we need to know that a suitable function exists to establish the identity (4.1). The actual form of the interpolating function g is not required, in general; again, it is sufficient to know that it exists, but generally reconstituting the waveform from its samples will not be required. In Sections 4.3 and 4.4 below (wideband and uniform sampling), we simply repeat the spectrum of u. In Section 4.5 (Hilbert sampling), we also include the spectrum of $\hat{u}$, the Hilbert transform of u and in Section 4.6 (quadrature sampling) we include a quarter wave delayed form of u. The sampling techniques of Sections 4.4 and 4.6 are for narrowband waveforms—signals on a carrier.

4.3 Wideband Sampling

By a wideband waveform u we mean a waveform containing energy at all frequencies from zero up to some maximum W beyond which there is no spectral energy. A real waveform has a complex spectrum U which is complex conjugate symmetric about zero, so real waveforms of interest have spectra within the interval $[-W, W]$ (Figure 4.1). If we repeat this spectrum at intervals $2W$, it will not overlap, as there is no spectral energy outside this interval, so we can write the identity

$$U(f) = \text{rep}_{2W}\, U(f)\, \text{rect}\,(f/2W) \tag{4.5}$$

where we have equated the spectrum to a gated portion of the repeated form of the spectrum itself (Figure 4.1). Taking the Fourier transform, we obtain

$$u(t) = \text{comb}_{1/2W}\, u(t) \otimes \text{sinc}\; 2Wt \tag{4.6}$$

This is the particular form of (4.2) for this sampling case. This equation states that u is equal to itself sampled at a rate of $2W$ (i.e., at intervals $T = 1/2W$) and correctly interpolated, the interpolating function in this case being $2W\,\text{sinc}\; 2Wt$. The equivalent form of (4.4) is

$$u(t) = \Sigma u(nT)\, \text{sinc}\,(t - nT) \tag{4.7}$$

and the equivalence of the waveform to its interpolated sampled form is illustrated in Figure 4.2.

It is clear (from Figure 4.1, for example) that if we repeat the spectrum at intervals $2W'$, where $W' > W$, we still obtain the spectrum U on gating with either $2W$ or $2W'$ bandwidth. Thus, any sampling rate greater than $2W$ is also adequate.

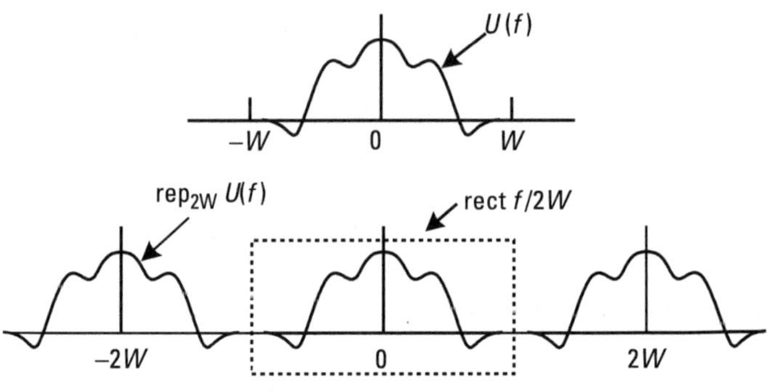

Figure 4.1 Gated repeated waveform.

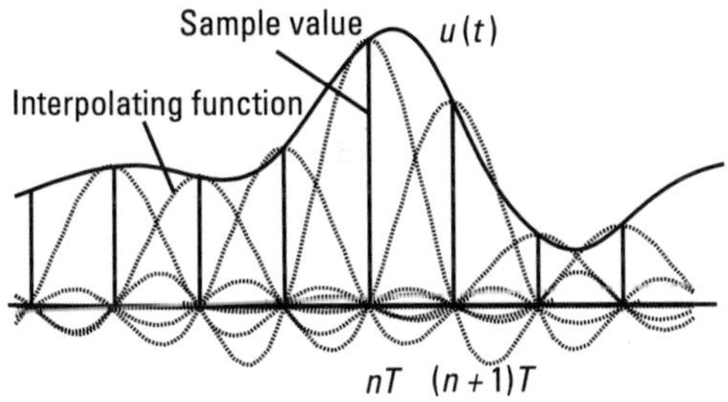

Figure 4.2 Sampled waveform with interpolating functions.

Thus we have the *wideband sampling theorem:*

> If a real waveform has no spectral energy above a maximum frequency W, then all the information in the waveform is retained by sampling it at a rate $2W$ (or higher).

In principle, reconstituting the waveform in this case is achieved by driving a rectangular bandwidth low-pass filter with impulses of strength proportional to the sample values and at the sample times. In practice, an approximation to u could be formed easily as a boxcar waveform from the sample values [simply holding the value $u(nT)$ constant over the interval $[nT, (n + 1)T]$. Smoothing this with a low-pass filter would give a better approximation to u.

4.4 Uniform Sampling

4.4.1 Minimum Sampling Rate

We define a real narrowband (or IF) waveform as one that has negligible power outside a frequency band W centered on a carrier frequency f_0, where $W/2 < f_0$. (The complex spectrum of a real IF waveform consists of two bands centered at $+f_0$ and $-f_0$. We label these U_+ and U_- as before, for convenience, as shown in Figure 4.3.) For such a waveform, it is not necessary to sample at twice the maximum frequency (i.e., at $2f_0 + W$ here), as in the case of a wideband waveform, but at approximately twice the bandwidth.

We initially restrict W so that the upper edge of the signal band f_u is an integer multiple of W, that is, $f_u = f_0 + W/2 = kW$ for k integral. The lower edge of the band is then at $(k - 1)W$. The spectrum can now be

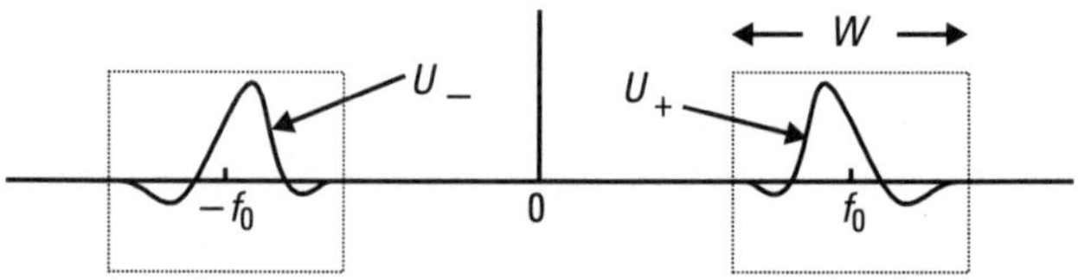

Figure 4.3 Narrowband spectrum.

repeated at intervals $2W$ without overlap as $2f_0 = (2k - 1)W$, so a displacement of $2kW$ or $2(k-1)W$ moves the spectral band U_-, centered at $-f_0$, adjacent to the band U_+ at f_0 without overlapping it (Figure 4.4). Thus, we can write

$$U(f) = \text{rep}_{2W}\, U(f)\left\{\text{rect}\left[\frac{(f - f_0)}{W}\right] + \text{rect}\left[\frac{(f + f_0)}{W}\right]\right\} \quad (4.8)$$

again representing U as a gated repeated form of itself. The transform of this equation is

$$u(t) = \frac{1}{2W}\,\text{comb}_{1/2W}\, u(t) \otimes W \,\text{sinc}\; Wt\left(e^{2\pi i f_0 t} + e^{-2\pi i f_0 t}\right) \quad (4.9)$$

Thus u is equal to itself sampled at a rate $2W$ and interpolated by the function sinc $Wt \cos(2\pi f_0 t)$, which is the impulse response of an ideal rectangular band-pass filter of bandwidth W centered at frequency f_0.

We now remove the condition relating f_0 and W. We note that a spectrum within a band $(f_u - W, f_u)$ is also within the band $(f_u - W', f_u)$ if $W' \geq W$. Thus, if W does not satisfy the condition $f_u = kW$ (k integral), we choose the smallest $W' > W$ that does satisfy it. More specifically, if $f_u = (k + \alpha)W$ where $0 \leq \alpha < 1$, we choose W' so that $f_u = kW'$, and we can write $k = [f_u/W]$, where $[x]$ means the largest integer contained in x. Then repeating the spectrum at intervals $2W'$ again produces a nonover-

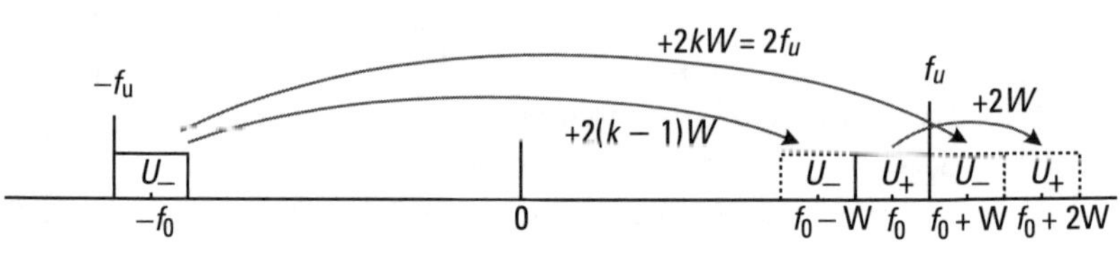

Figure 4.4 Allowed spectral shifts.

lapping spectrum (Figure 4.5), but this time with some gaps [of size $2(W' - W)$] due to the difference between W' and W.

It is clear that to regain U_+ from the part of the spectrum shown in Figure 4.5, it is only necessary to gate with the same gating function as before [given in (4.8)—gates of width W centered at $+f_0$ and $-f_0$] leading to the same interpolating function, sinc $Wt \cos(2\pi f_0 t)$. Brown [2] in effect uses the more complicated interpolating function $\text{sinc}(2W't)\cos 2\pi f_0' t$ where $f_0' = f_0 - (W' - W)/2$. This corresponds to using the gating function rect $[(f - f_0')/W']$, which will also gate out U_+ as required (Figure 4.6) but is more complicated than necessary.

4.4.2 General Sampling Rate

The minimum sampling rate $2f_u/k$, found in the previous section, is such that the band U_- shifted by $2kW'$ is just above U_+ when the repetitive spectrum is formed (Figure 4.5). If W' is increased above this value, this band will move up in frequency, and so will the band U_-, shifted by $2(k - 1)W'$, which will eventually start to overlap U_+. This will define a (local) maximum allowed sampling rate, and this occurs when $2(k - 1)W' = 2f_l$, where f_l is the frequency at the lower edge of the signal band (Figure 4.7). Thus, the allowed sampling rate $2W'$ ranges from a minimum value $2f_u/k$ to a maximum $2f_l/(k - 1)$. As k is defined here by $f_u = (k + \alpha)W$, we also have $f_l = f_u - W = (k - 1 + \alpha)W$, and we see that the range of allowed sampling rates $2W'$ is given by

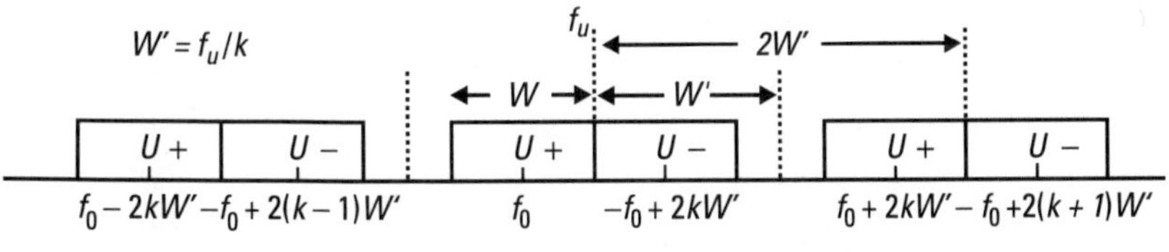

Figure 4.5 rep$_{2W}$ $U(f)$ near $+f_0$.

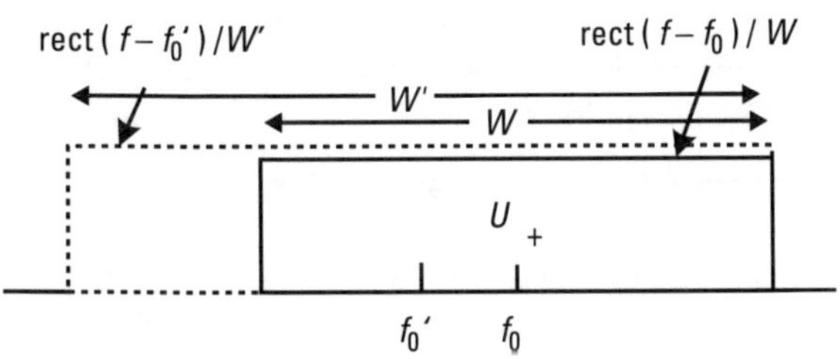

Figure 4.6 Selecting $U(f)$.

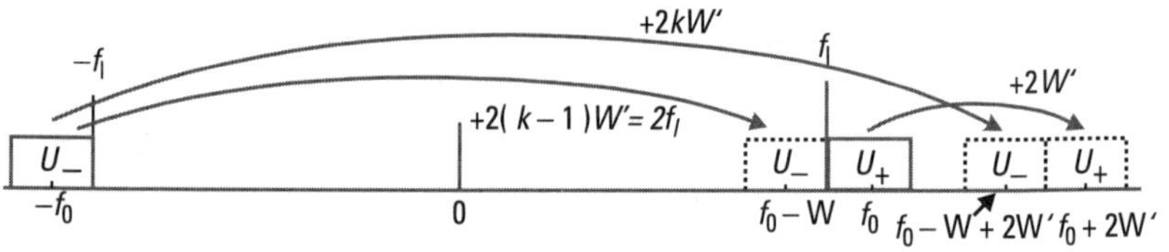

Figure 4.7 Maximum sampling rate.

$$2f_u/k = 2(k + \alpha)\,W/k \leq 2W' \leq 2(k - 1 + \alpha)\,W/(k - 1) = 2f_l/(k - 1) \tag{4.10}$$

It is convenient to define a relative sampling rate r as the actual rate divided by the minimum value possible (to retain all the signal data) $2W$, so that the allowed relative rate $2W'/2W$ becomes

$$(k + \alpha)/k \leq r \leq (k - 1 + \alpha)/(k - 1) \tag{4.11}$$

or

$$1 + \alpha/k \leq r \leq 1 + \alpha/(k - 1) \tag{4.12}$$

If the sampling rate is increased above the "maximum" $2f_l/(k - 1)$, we see from Figure 4.7 that U_- will overlap U_+ until the rate rises to $2f_u/(k - 1)$ when we reach a new local minimum value for the allowed sampling rate. The rate can now be increased to a new local maximum $2f_l/(k - 2)$ before overlap starts again. In general, we see that allowed relative sampling rates are given by

$$(k + \alpha)/n \leq r \leq (k - 1 + \alpha)/(n - 1) \qquad (n = k,\, k - 1,\, \ldots,\, 1) \tag{4.13}$$

(In the $n = 1$ case, we only have a minimum rate; the maximum rate in this case is unbounded.) Putting $n = k$ gives the absolute minimum rate, $1 + \alpha/k$. The allowed relative sampling rates are given in the shaded regions of Figure 4.8 as a (multivalued) function of the center frequency normalized to the bandwidth.

We note from Figure 4.8 that the lowest range of allowed rates becomes very narrow at high values of f_0/W. This indicates that the sampling rate should be carefully chosen in this case, and perhaps should be synchronized to some frequency in the signal band. The minimum rate is in fact defined by f_u, but there is no actual signal power here (from the definition of W),

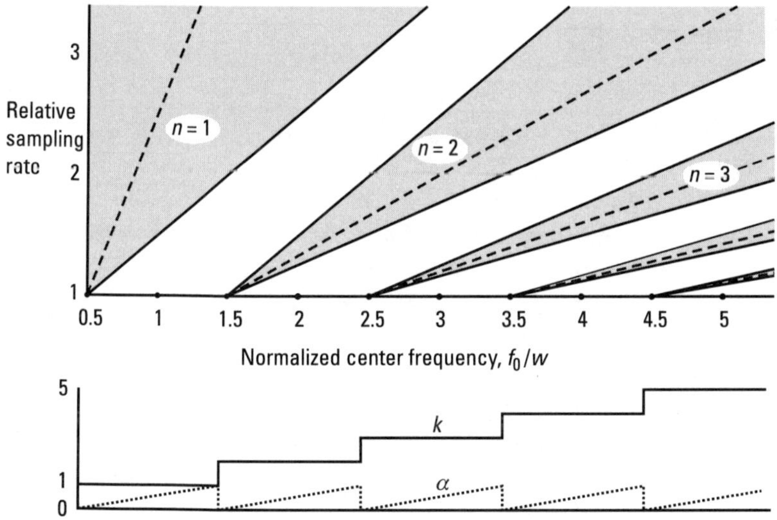

Figure 4.8 Relative sampling rates (uniform sampling).

so it would be more convenient to use f_0. The allowed band of relative rates, from (4.12), is between $1 + \alpha/k$ and $1 + \alpha/(k - 1)$, so $1 + \alpha/(k - ½)$ would be near the mean of these. The actual rate with this choice is thus $2W(k + \alpha - ½)/(k - ½) = 2(f_u - W/2)/(k - ½) = 2f_0/(k - ½)$. This rate is indicated by the dashed lines and is very close to the minimum rate for higher values of f_0/W (e.g., above 3½).

We note that if $f_0/W = ½$, we have effectively a wideband waveform of positive frequency bandwidth W (see Figure 4.3), with positive frequencies extending from 0 to W. The uniform sampling rate in this case is $2W$ (see Figure 4.8), which agrees with the result in Section 4.3 for a wideband waveform. We also note that the minimum rate is substantially different from $2W$ only for large fractional bandwidths (W/f_0 large and so f_0/W small). For small fractional bandwidths of, for example, a few percent, which is often the case for radio and radar signals, whether at high, very high, ultra-high, or microwave frequencies, the correct rate will be very close to $2W$, and setting it actually at $2W$ will generally give negligible degradation.

Finally, we state a simplified form of the *uniform sampling theorem* for narrowband waveforms, which is not as neatly defined as for the wideband case:

> If a real waveform has no spectral energy outside a frequency band of width W centered on a carrier of frequency $f_0 = (k - ½ + \alpha)W$ (k integral, $0 \le \alpha < 1$), then all the information in the waveform is retained by sampling it at a rate $2rW$, where r is given in (4.13) above.

4.5 Hilbert Sampling

Given a real waveform u, the complex waveform $v = u + i\hat{u}$ has a spectrum consisting of positive frequency components only, where $\hat{u}$ is the Hilbert transform of u, defined in Appendix 4A. For narrowband waveforms, a 3-dB coupled line directional coupler is a very good approximation to a Hilbert transformer, which generates $\hat{u}$ from u. (The Hilbert transformer effectively produces a wideband 90-degree phase shift; see Appendix 4A.) The two outputs of such a coupler are a Hilbert transform pair, which may be considered to form a complex waveform, if the rules for complex arithmetic are observed when processing this two-channel waveform.

If W is the band (centered on $-f_0$ and f_0) outside which U has negligible power, then we can see that V is within a band of width W, centered on $+f_0$ only, and so can be repeated at intervals W without overlapping (Figure 4.9). Thus, we can write the identity

$$V(f) = \text{rep}_W\, V(f)\, \text{rect}\,[(f - f_0)/W] \tag{4.14}$$

Performing the inverse Fourier transform using P3b, R8a, R6b, and R5, we obtain

$$v(t) = (1/W)\,\text{comb}_{1/W}\, v(t) \otimes [W\,\text{sinc}\; Wt\, \exp(2\pi i f_0 t)] \tag{4.15}$$

Now u is the real part of v, so taking the real part of both sides we obtain

$$u(t) = \text{comb}_{1/W}\, u(t) \otimes (\text{sinc}\; Wt \cos 2\pi f_0 t) - \text{comb}_{1/W}\, \hat{u}(t) \otimes (\text{sinc}\; Wt \sin 2\pi f_0 t) \tag{4.16}$$

We see that u is equal to a combination of samples of u and $\hat{u}$ appropriately interpolated, the samples being taken at intervals $1/W$ (i.e., at a rate W). We also note that by taking the imaginary part of v from (4.15), we obtain the sampled form of $\hat{u}$. If we repeat the spectrum at intervals $W' > W$, corresponding to sampling at the rate W', we still have a nonoverlap-

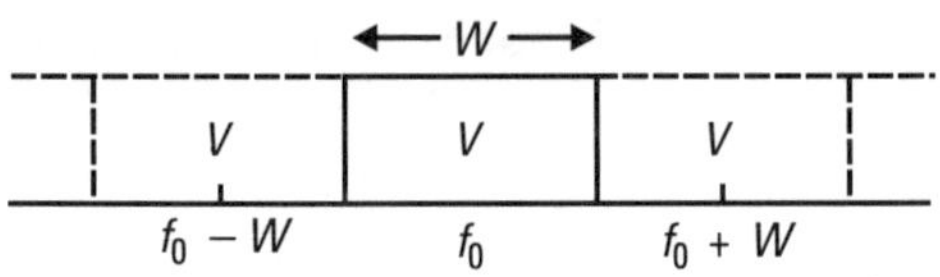

Figure 4.9 $\text{rep}_W\, V(f)$ near $+f_0$.

ping spectrum. This could be gated with a rectangular window of any width from W to W' to obtain V again. Thus we obtain the *Hilbert sampling theorem,* which is more simply stated than the uniform sampling theorem for the same type of waveform:

> If a real waveform u has no spectral energy outside a frequency band of width W centered on a carrier of frequency f_0, then all the information in the waveform is retained by sampling it and its Hilbert transform $\hat{u}$ at a rate W. The samples are complex, the real parts are the samples of u, and the imaginary parts are the samples of $\hat{u}$.

We note that the sampling rate is independent of f_0, unlike the case for uniform sampling or quadrature sampling (an approximation to Hilbert sampling, described in Section 4.6 below). As pointed out by Woodward, a real waveform of duration T and bandwidth W requires (as a minimum) $2WT$ real values to specify it completely—either real samples at a rate $2W$ (as given by wideband sampling, or as the minimum rate in the case of uniform or quadrature sampling) or WT complex samples (containing WT real values in each of the real and imaginary parts) in the case of Hilbert sampling. The waveform can be said to require $2WT$ degrees of freedom for its specification.

4.6 Quadrature Sampling

4.6.1 Basic Analysis

If it is not convenient or practical to use a quadrature coupler or any other method to produce the Hilbert transform of a narrowband waveform, an approximation to the transformed waveform can be obtained by delaying the signal by a quarter cycle of its carrier frequency. This follows from the fact that the Hilbert transform is equivalent to a delay of $\pi/2$ radians (for all frequency components, as shown in Appendix 4A), so the quarter cycle delay will be correct at the center frequency and nearly so for frequencies close to it. The smaller the fractional bandwidth, the better this approximation becomes. As this is an approximation to the Hilbert transform, it follows that sampling at the rate $2W$ (the Hilbert sampling rate) will not, in general, sample the waveform adequately (to retain all the information contained in it). However, we will see, below, that the method will in fact sample correctly, but at the cost, compared with Hilbert sampling, of requiring an increased sampling rate, the rate depending on the ratio of bandwidth to center frequency (similar to the case of uniform sampling).

If $u(t)$ is the basic waveform, with spectrum $U(f)$, then a delayed version $u(t-\tau)$ has spectrum $U(f)\exp(-2\pi if\tau)$. If we repeat the spectrum of u at intervals W, corresponding to sampling at the rate W, we will obtain an overlapping spectrum that, when gated, is not equal to U in general. However, a suitable combination of the repeated spectra of u and its delayed version will give U after gating. We start by imposing the condition $2f_0 = kW$, where k is an integer, so that there is complete overlap of the two parts of the spectrum of u, and also of the two parts of the spectrum of its delayed version when repeated (Figure 4.10).

The appropriate identity for U is

$$U(f) = \frac{1}{2}\{\text{rep}_W U(f) + \exp(2\pi if\tau)\,\text{rep}_W[U(f)\exp(-2\pi if\tau)]\} \quad (4.17)$$
$$\cdot\,[\text{rect}\,(f-f_0)/W + \text{rect}\,(f+f_0)/W]$$

if τ is correctly chosen. To check this identity, we consider the output of the positive frequency spectral gate for frequencies in the range $f_0 - W/2 < f < f_0 + W/2$. In this interval we have, as there is overlap of the negative frequency part of the spectrum, moved up by $2f_0$, or kW, for some integer k,

$$\frac{1}{2}\{U(f) + U(f-2f_0) + \exp(2\pi if\tau)[U(f)\exp(-2\pi if\tau)$$
$$+ U(f-2f_0)\exp(-2\pi i(f-2f_0)\tau)]\} \quad (4.18)$$
$$= U(f) + \frac{1}{2}U(f-2f_0)\,[1 + \exp(4\pi if_0\tau)]$$
$$(f_0 - W/2 < f < f_0 + W/2)$$

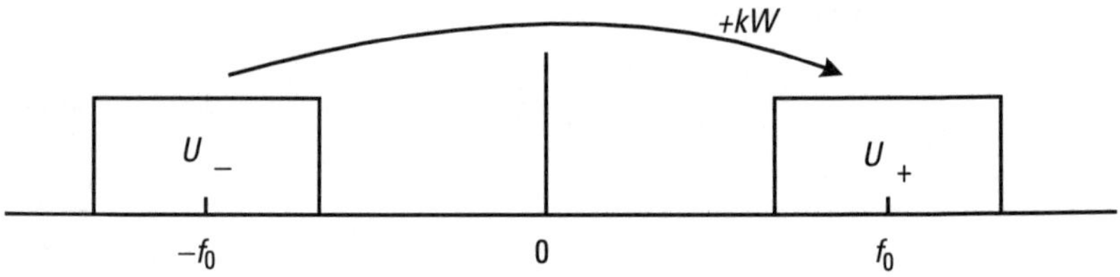

Figure 4.10 Basic quadrature sampling.

This is simply $U(f)$, as required, if we choose τ such that $4f_0\tau = 1$, or, more generally, if $4f_0\tau = 2m + 1$, where m is an integer. The same condition results if we consider the output of the negative frequency gate—we simply replace f_0 with $-f_0$ throughout. Thus, the required delay is seen to be an odd number of quarter wavelengths of the carrier, or center frequency f_0, or one quarter cycle in the simplest case. Taking the (inverse) Fourier transform of the identity for $U(f)$ in (4.17), we have

$$\begin{aligned} u(t) &= \frac{1}{2}\,[(1/W)\ \text{comb}_{1/W}\,u(t) + \delta(t+\tau) \\ &\quad \otimes (1/W)\ \text{comb}_{1/W}\,u(t-\tau)] \otimes 2W\phi(t) \qquad (4.19) \\ &= \text{comb}_{1/W}\,u(t) \otimes \phi(t) + \text{comb}_{1/W}\,u(t-\tau) \otimes \phi(t+\tau) \end{aligned}$$

where ϕ is the interpolating function. This is obtained from the (inverse) Fourier transform of the spectral gating function Φ, defined by

$$2W\Phi(f) = \text{rect}\,[(f - f_0)/W] + \text{rect}\,[(f + f_0)/W] \qquad (4.20)$$

Thus,

$$2W\phi(t) = W\,\text{sinc}\,(Wt)\,[\exp\,(2\pi i f_0 t) + \exp\,(-2\pi i f_0 t)]$$

or

$$\phi(t) = \text{sinc}\,(Wt)\cos\,(2\pi f_0 t) \qquad (4.21)$$

This interpolating function also appears in the uniform sampling case [see (4.9)] and the Hilbert sampling case [see (4.16)]. Equation (4.19) states that the real waveform u is equal to the sum of the waveform obtained by sampling u at intervals $1/W$ (i.e., at rate W) and interpolating with the function ϕ and the waveform obtained by sampling a quarter-wave delayed version of u and interpolating with a quarter-wave advanced version of ϕ.

To remove the condition relating W and f_0, we choose $W' \geq W$ such that $2f_0 = kW'$, where $k = [2f_0/W]$, the largest integer in $2f_0/W$. We then repeat the spectrum at intervals W', which corresponds to sampling at the rate W', but we can keep the same spectral gating function and hence the same interpolating function. The minimum required sampling rate, relative to the minimum rate, equal to the bandwidth W, is $r = W'/W = 1 + \alpha/k$

if $2f_0/W = k + \alpha$. This minimum rate is plotted in Figure 4.11, and this is the rate given by Brown [2].

If W' is increased to higher values such that $2f_0 = nW'$ for n integral, $n < k$, we again obtain sampling rates that will retain the waveform information, and these are shown by the dashed lines in Figure 4.11. The required sampling frequency could be obtained in practice by synchronizing W' to a submultiple of $2f_0$ (ideally the kth, for the minimum rate).

4.6.2 General Sampling Rate

Unlike the uniform sampling case, the required sampling rates determined so far are precise (Figure 4.11) instead of within bands (as in Figure 4.8). This is because the delay has been chosen to be a quarter cycle of f_0 (or an odd number of quarter cycles). In fact, on replacing $2f_0$ with kW' in (4.18), where kW' is the frequency shift that takes U_-, centered at $-f_0$, onto U_+, centered at $+f_0$, we see that the condition to be satisfied is $2kW'\tau = 2m + 1$ (m an integer). If we relate the delay τ to the sampling rate W' instead of directly to f_0, then we have more freedom of choice of W'. In Figure 4.12(a), we see part of the function $\text{rep}_W\, U_-$, the signal band at $-f_0$ repeated at intervals W, in the region of $+f_0$ where $2f_0$ is not an integer multiple of W. If we consider the part of this spectrum that overlaps the band of width W, centered at $+f_0$, we see that there is a mixture of parts of U_- shifted by kW and by $(k + 1)W$. If the delay is correct to make U_- disappear when shifted by kW, then it is not quite correct when shifted by $(k + 1)W$, and a small amount of spectral overlap occurs.

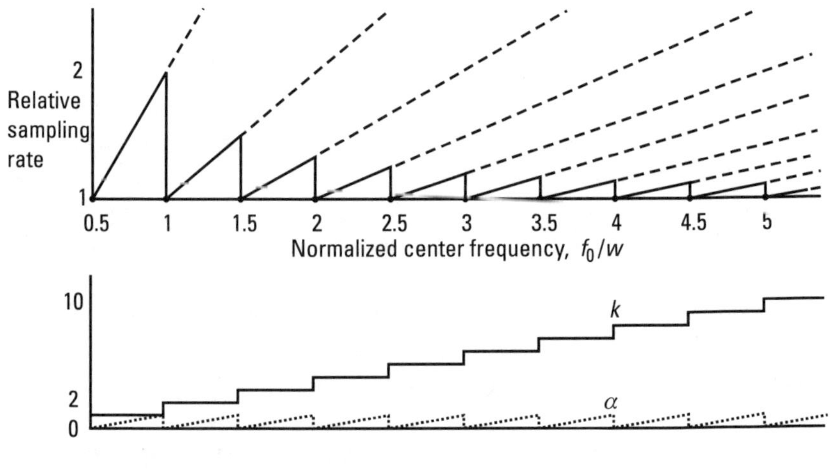

Figure 4.11 Relative sampling rates (basic quadrature sampling).

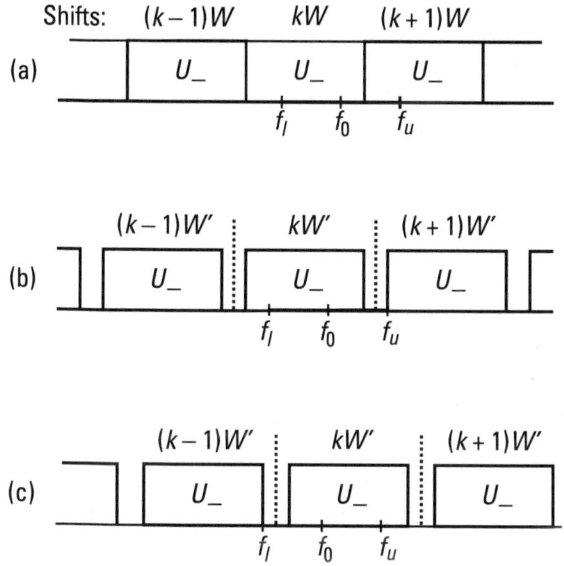

Figure 4.12 Shifted positions of U_-: (a) $2f_0 = (k + \alpha)W$, $(0 < \alpha < 1)$; (b) $2f_u = (k + 1)W'$; and (c) $2f_1 = (k - 1)W'$.

The minimum repetition rate to avoid this is shown in Figure 4.12(b), where W' ($> W$) is such that $(k + 1)W'$ moves U_- just beyond the gated region (between f_1 and f_u). Because $W' > W$ gaps of width $W' - W$ now occur between the repeated versions of U_-. The minimum required value of W' is given by $(k + 1)W' = 2f_u$. (In fact, other local minimum rates are given by W' such that $(n + 1)W' = 2f_u$, for n integral $n < k$, but we will see that we do not need to consider these rates because of a more general result below.) In this case, in order for U_- to disappear in the gated band, the delay must satisfy $2kW'\tau = 1$, and so, with the condition on W' above, we find that $\tau = (k + 1)/4kf_u$—that is, the delay should be $(1 + 1/k)$ times a quarter cycle of the upper edge of the signal band f_u (or an odd multiple of this).

If we increase the sampling rate further, we reach the condition shown in Figure 4.12(c), where the band U_- has just reached the lower edge of the gated band. This is when $(k - 1)W' = 2f_l$ (or, again, more generally when $(n - 1)W' = 2f_l$ for n an integer and $n \leq k$). The delay required is $(1 - 1/k)$ times a quarter cycle of the lower edge of the signal band f_l (or an odd multiple of this).

To summarize, the minimum and maximum relative sampling rates are given by $r_m = 2f_u/W(n + 1)$ and $r_M = 2f_l/W(n - 1)$, where $f_u = f_0 +$

$W/2$ and $f_l = f_0 - W/2$; a central rate (very close to the mean of these two) is $r_c = 2f_0/nW$. With k and α defined by $2f_0 = k + \alpha$, these become $r_m = (k + \alpha + 1)/(n + 1)$, $r_M = (k + \alpha - 1)/(n - 1)$, and $r_c = (k + \alpha)/n$, where $n \le k$ (n and k integers and $0 \le \alpha < 1$). When $n = k$, these become $r_m = 1 + \alpha/(k + 1)$, $r_M = 1 + \alpha/(k - 1)$, and $r_c = 1 + \alpha/k$. Also, when $\alpha = 0$ ($2f_0/W$ integral), then these rates are all unity, and $n < k$ corresponds to the continuations of these particular lines from lower k values, as illustrated in Figure 4.13.

The allowed sampling rates relative to the bandwidth W are given in the shaded areas in Figure 4.14. The maximum and minimum rates r_M and r_m define the boundaries, and the central values rc are shown as dashed lines in Figure 4.14. We note from Figure 4.14 that there are no unallowed sampling rates above $2W$. This is because when the interval between repetitions of U_- becomes $2W$, it is not possible to have parts of more than one repetition of U_- in the gating interval (see Figure 4.12(b or c) with $W' \ge 2W$), so if the delay is correctly chosen, the U_- contribution in this interval can always be removed. [By putting $x = f_0/W = (k + \alpha)/2$ and equating r_m at $n = k - 1$ and r_m at $n = k$, with $\alpha = 2x - k$, we find these lines meet at $x = k - \frac{1}{2}$ and the common value of r is 2, as shown in Figure 4.14.]

However, the general rates given in Figure 4.14 may not be very convenient in practice, as they require choosing the delay to be $1/2kW'$ [where W' is between $2f_u/(k + 1)$ and $2f_1/(k - 1)$], which may not be as easy as choosing it to be $1/4f_0$, as assumed in Figure 4.11. Because the required delay is no longer exactly a quarter cycle (or an odd number of quarter cycles) of the carrier, this sampling has been termed *modified quadrature sampling* in the title of Figure 4.14. In fact, the central rate $2f_0/k$ does require the quarter cycle delay, but from this study we see that this is not the minimum rate, if that is what is required.

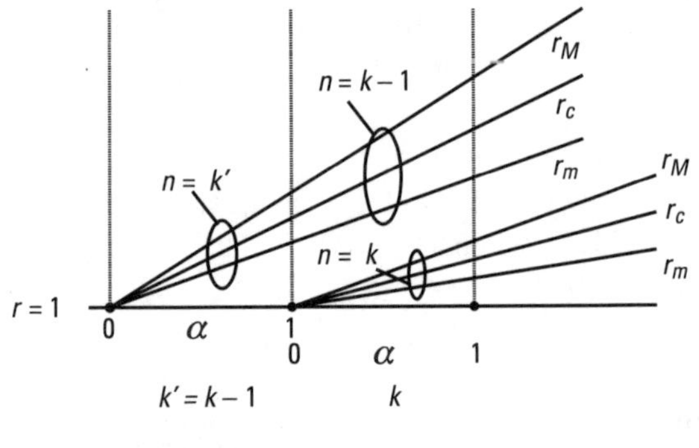

Figure 4.13 Lines of relative sampling rates.

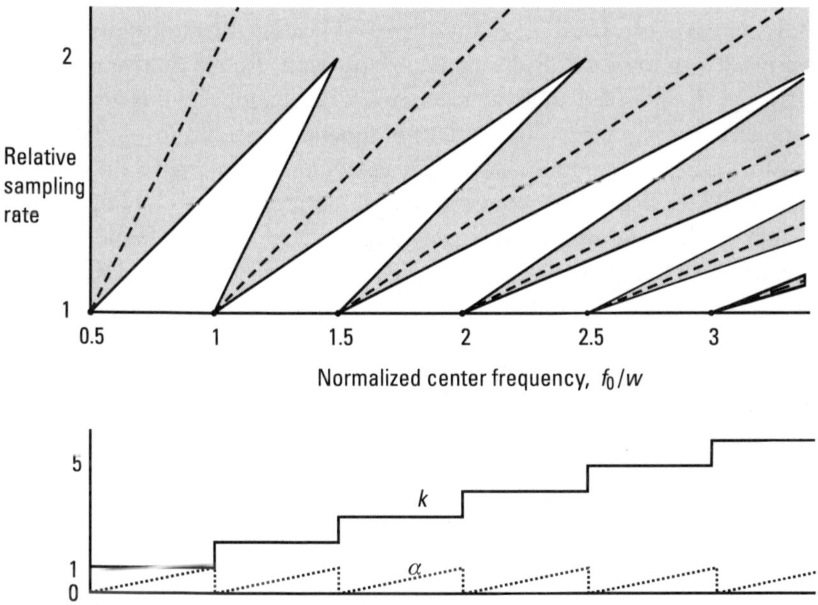

Figure 4.14 Relative sampling rates (modified quadrature sampling).

Thus, we can now state a quadrature sampling theorem:

> If a real waveform u has no spectral energy outside a frequency band of width W centered on a carrier of frequency f_0, then all the information in the waveform is retained by sampling it and a delayed version of it at a rate given by rW, where r is given in (4.22) and the delay (which is close to a quarter cycle of f_0) is given in (4.23). The samples are complex, the real parts are the samples of u, and the imaginary parts are the samples of the delayed form.

4.7 Low IF Analytic Signal Sampling

A signal $u(t)$ on a carrier at frequency f_0 can be written $u(t) = a(t)\cos[2\pi f_0 t + \phi(t)]$, and, at least in principle, we can derive its Hilbert transform, $\hat{u}(t) = a(t)\sin[2\pi f_0 t + \phi(t)]$, and hence the complex form $u(t) + i\hat{u}(t) = a(t)\exp i[2\pi f_0 t + \phi(t)]$. The information in this signal is contained in the amplitude and phase functions $a(t)$ and $\phi(t)$, and what is required for digital signal processing is a digital form of the analytic signal $a(t)\exp i\phi(t)$. This is what is given by Hilbert sampling and quadrature sampling, discussed above, in particular from the point of view of finding the minimum sampling rate needed to preserve all the signal information. An alternative method of obtaining the sampled analytic, or complex baseband, signal is given in this section. This is simpler to implement in practice—not requiring the Hilbert transform or an accurate quarter cycle delay, and sampling in only a single channel rather than in two—at the cost of requiring a higher sampling rate, though, at the minimum, this single sampling device [or analogue-to-digital converter (ADC)] operates at just twice the rate of the two ADCs needed for the alternative methods.

The method requires bringing the signal carrier frequency down from the normally relatively high RF to a low IF. To avoid the two parts of the spectrum overlapping, we see that we must have $f_0 \geq W/2$. The samples we require are those corresponding to the complex baseband waveform $V(f)$, given by

$$V(f) = 2U_+(f + f_0) \tag{4.22}$$

which is the positive frequency part of the spectrum (the spectrum of the equivalent complex waveform) centered at zero frequency (baseband) rather than at the IF carrier f_0. We see that, given U, we can obtain V by first shifting U by $-f_0$, then gating it with $2\,\mathrm{rect}\,(f/W)$ (Figure 4.15). In order to obtain the repetitive element in the spectrum, to give the δ-functions in the time domain corresponding to the sample values, we note that we can repeat this shifted U spectrum, without overlapping, at intervals $F \geq 2f_0 + W$, so that we have

$$V(f) = \mathrm{rep}_F\,[2U(f + f_0)]\;\mathrm{rect}\,\frac{f}{W} \tag{4.23}$$

Taking the (inverse) transform, using P3b, R8a, R6b, and R7b we have

$$v(t) = \frac{1}{F}\,\mathrm{comb}_{1/F}\,[2u(t)\exp(-2\pi i f_0 t)] \otimes W\,\mathrm{sinc}\,Wt \tag{4.24}$$

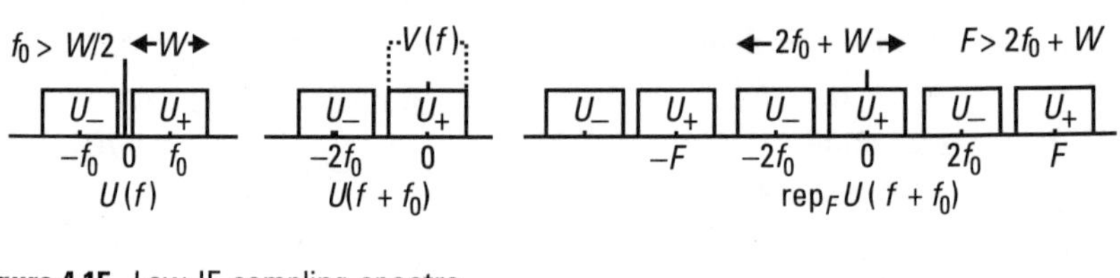

Figure 4.15 Low IF sampling spectra.

or

$$v(t) = \frac{2W}{F} \operatorname{comb}_T [u(t) \exp(-2\pi i f_0 t)] \otimes \operatorname{sinc} Wt \qquad (T = 1/F)$$

Thus the analytic, complex baseband waveform is given by sampling the real IF waveform u multiplied by the complex exponential $\exp(-2\pi i f_0 t)$—that is, mixing down to baseband using a complex local oscillator (LO) at the signal's center frequency f_0. (Again, in principle, to form this waveform, we interpolate the samples obtained at intervals $T = 1/F$, where the sampling rate F is $2f_0 + W$ or higher, with sinc functions.) In fact, we do not have to provide this LO waveform in continuous form, as we note that

$$\operatorname{comb}_T [u(t) \exp(2\pi i f_0 t)] = \sum_{n=-\infty}^{\infty} u(nT) \exp(-2\pi i n f_0 T)\, \delta(t - nT) \tag{4.25}$$

and we see that we multiply the samples of u by the sampled form of the complex exponential. In the case where the IF carrier is $f_0 = W/2$ and the sampling rate F is the minimum $2W$, we see that F is just $4f_0$ and the sampled complex LO values are given by $\exp(-\pi i n/2)$ or $(-i)^n$—that is, we just multiply the real samples of u by $1, -i, -1$, and i in turn, a particularly simple form of down-conversion.

If the IF is greater than $W/2$ (up to $3W/2$), then we can repeat the spectrum at the smaller interval of $2f_0 + W$, rather than $4f_0$, but in this case the complex down-conversion factors are not so simple (being given by $\exp[-\pi i n/(1 + W/2f_0)]$) and generally the $4f_0$ sampling rate will be preferred. If the IF is considerably higher than the bandwidth, then lower sampling rates that avoid overlapping can be used, as discussed in Section 4.4, under the topic of uniform sampling, of which this method is an example. Using the notation of Section 4.4, the lowest IF case corresponds to $f_u = W$ and $k = 1$. For higher IF values we have $f_u = f_0 + W/2 = kW'$, where W' is the lowest value above (or equal to) W such that f_u/W' is an integer k. Then the minimum required sampling rate is $2W' = (2f_0 + W)/k$, and the complex down-conversion factors are $\exp(-2\pi i f_0 nT)$, where $T = 1/2W'$, leading to the factors $\exp[-\pi i k n/(1 + W/2f_0)]$. Again, this is an awkward form to apply, but if we chose the slightly higher sampling rate of $2f_0/(k - ½)$, as suggested in Section 4.4, then the down-conversion factors become simply $\exp[-\pi i n(k - ½)]$ or $-i^n$ for k odd and i^n for k even. However, sampling with a finite window width on a high IF may require care, as discussed in the next section, and keeping the IF low would generally be preferable.

4.8 High IF Sampling

If we sample at a relatively high IF, the time taken to obtain a sample of the waveform may become significant compared with the period of the carrier. We take for our model a device that integrates the waveform over a short interval τ, the sample value recorded being the mean waveform value over this interval, the integral divided by τ. We see that this value is the same as would be given by a device that sampled instantaneously the waveform given by sliding a $(1/\tau)\,\mathrm{rect}\,(t/\tau)$ function across the waveform and integrating—that is, forming the convolution of the waveform with the rect function. Thus, if u is the waveform, the samples actually correspond to the waveform v given by

$$v(t) = u(t) \otimes (1/\tau)\,\mathrm{rect}\,(t/\tau) \tag{4.26}$$

The spectrum of this is

$$V(f) = U(f) \cdot \mathrm{sinc}\,(f\tau) \tag{4.27}$$

Figure 4.16 shows the spectrum of V compared with that of U, shown as a rectangular band (in the positive frequency region only). With a low

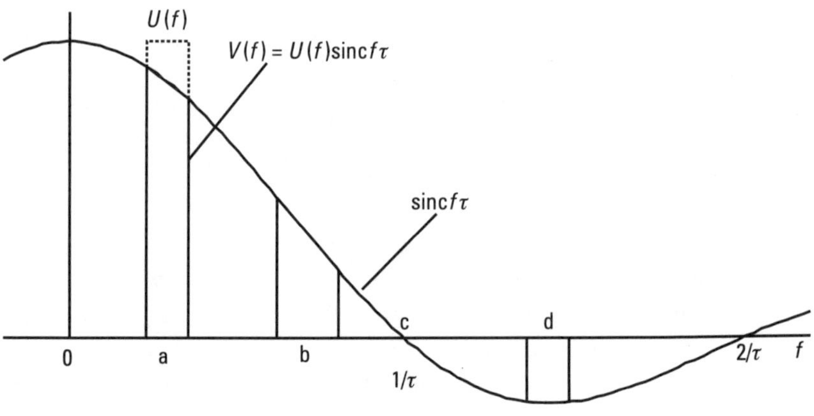

Figure 4.16 Spectrum of waveform sampled with a finite window.

carrier frequency f_0, compared with $1/\tau$ (i.e., with τ a small fraction of the period of the carrier), in position "a," there is relatively modest distortion across the signal band. At a higher center frequency, position "b," and also with a larger bandwidth, the distortion is more serious. At position "c," where the window is one cycle of f_0 ($f_0\tau = 1$), the distortion is severe and totally unacceptable. However, in position "d," where the sinc function is near a stationary value, the distortion is very low. This is at $f_0\tau = 1.434$, so the window τ should be about 1.4 cycles of the carrier for a low distortion result.

4.9 Summary

In this chapter we have shown how the rules-and-pairs method can be used to obtain some sampling results very neatly and concisely. The main aim was to determine the minimum sampling rates that would retain the signal information, but in some cases the method was used to find what other rates would be acceptable (not necessarily all rates above the minimum). This was first applied to sampling wideband signals, with significant spectral power from some maximum W down to zero frequency. The information in a real waveform is all retained by sampling it at the rate $2W$ (or any higher rate). The second example, uniform sampling, applies to a narrowband signal, a signal on a carrier, with a spectrum limited to a band of frequencies around the carrier. In this case, the rates acceptable are dependent on the ratio of the bandwidth W to the center frequency f_0 being at least $2W$ and generally higher. This form of sampling is an example of the case where some higher sampling rates are not allowed if distortion is to be avoided.

A different approach is to convert the real waveform into the complex waveform that has the given waveform as its real part. This requires deriving the imaginary part from the real part by means of a Hilbert transform. In principle, this is applicable to both wideband and narrowband waveforms, though it is more likely to be applied to the latter in practice. Given the complex waveform, we find very quickly that we only have to sample (in the two channels, real and imaginary) at the rate W (or any higher rate) to obtain complex samples representing the waveform. It is this complex form that is normally required for digital signal processing.

Hilbert sampling seems a very satisfactory approach, but it does depend on the provision of a good Hilbert transform, which is equivalent to a wideband (all-frequency) phase shift of 90 degrees. A close approximation to Hilbert sampling for narrowband waveforms is quadrature sampling, where the Hilbert transform is replaced by a delay essentially equal to a quarter of the carrier period. This provides the 90-degree shift of the carrier and close to 90 degrees for frequencies close to the carrier. However, it is not exact—the signal envelope is delayed in the imaginary channel, which is a form of distortion. Nevertheless, the analysis shows that all the data in the signal can be retained by sampling at the correct rate and with the correct delay, but generally this rate is higher than for Hilbert sampling, and, as for uniform sampling, depends on the ratio of W to f_0. Also, like the uniform sampling case, not all rates above the minimum are allowed.

The last method we consider is uniform sampling on a low IF, with down-conversion effectively achieved after sampling. This includes the case of sampling at four times the carrier (IF) frequency and gives a particularly simple way of providing the complex baseband samples without the need for a Hilbert transformer or a quarter wave delay, so it is an attractive method to implement. The required sampling rate in a single channel is, at the minimum, twice that needed in the two channels for the other methods.

Finally, we consider the effect of trying to sample on too high an IF. If the sampling gate duration becomes a significant fraction of the carrier period, then there will be some spectral distortion. This is very easily shown using a simple model for the sampling ADC. However, it is also shown that the spectral distortion can be made low by careful choice of the ratio of the high IF period to the sampling gate width.

References

[1] Woodward, P. M., *Probability and Information Theory: With Applications to Radar,* Norwood, MA: Artech House, 1980.

[2] Brown, J. L., Jr., "On Quadrature Sampling of Bandpass Signals," *IEEE Trans.,* AES-15, No. 3, 1979, pp. 366–371.

Appendix 4A: The Hilbert Transform

A real waveform u has a spectrum U of positive and negative frequencies, with all the information about it contained in one half of the spectrum. (We have already seen in Section 2.3 that the negative frequency components are just the complex conjugate of those of the corresponding positive frequencies.) We can define a complex function $v = u + i\hat{u}$ which has a positive frequency spectrum only, if we can form $\hat{u}$, with spectrum $\hat{U}$, such that $i\hat{U}$

is equal to U for positive frequencies and to $-\hat{U}$ for negative frequencies. Thus, given

$$v(t) = u(t) + i\hat{u}(t) \tag{4A.1}$$

with spectrum

$$V(f) = U(f) + i\hat{U}(f) \tag{4A.2}$$

then if we choose

$$i\hat{U}(f) = \begin{cases} U(f) & \text{for } f > 0 \\ -U(f) & \text{for } f < 0 \end{cases} \tag{4A.3}$$

(and $\hat{U}(0) = 0$) the spectrum of v is given by

$$V(f) = \begin{cases} 2U(f) & \text{for } f > 0 \\ 0 & \text{for } f < 0 \end{cases} \tag{4A.4}$$

[and $V(0) = U(0)$]. This is a spectrum of positive frequencies only, as required. To find $\hat{u}$, we note from (4A.4) that $V(f)$ can be written as $2U(f)h(f)$; so taking the inverse transform, using P2b, we have

$$v(t) = 2u(t) \otimes \left[\frac{\delta(t)}{2} - \frac{1}{2\pi it}\right] = u(t) + iu(t) \otimes \left(\frac{1}{\pi t}\right)$$

and so

$$\hat{u}(t) = u(t) \otimes \left(\frac{1}{\pi t}\right) = \frac{1}{\pi}\int_{-\infty}^{\infty} \frac{u(\tau)}{t - \tau}\, d\tau \tag{4A.5}$$

The Hilbert transform of $u(t) = \cos 2\pi f_0 t$ is $\hat{u}(t) = \sin 2\pi f_0 t$; this can be found using (4A.5) (treating τ as a complex variable and using contour integration) or, more simply, by choosing the function for $\hat{u}$ that converts the two-line spectrum (at $-f_0$ and $+f_0$) of u into the single-line spectrum of v (at $+f_0$ only)—that is, which makes v a single complex exponential. In this case, $v(t)$ is given by

$$v(t) = \cos 2\pi f_0 t + i \sin 2\pi f_0 t = \exp 2\pi i f_0 t$$

and so $V(f) = \delta(f - f_0)$, which is a single line at $+f_0$. The spectra of u and $i\hat{u}$ are $\frac{1}{2}[\delta(f - f_0) + \delta(f + f_0)]$ and $\frac{1}{2}[\delta(f - f_0) - \delta(f + f_0)]$, respectively, which satisfy the form of (4A.3) above. Similarly, the Hilbert transform of $\sin 2\pi f_0 t$ is $-\cos 2\pi f_0 t$, so that in this case,

$$v(t) = \sin 2\pi f_0 t - i \cos 2\pi f_0 t = -i \exp 2\pi i f_0 t$$

Both these Hilbert transforms correspond to a phase shift of $-\pi/2$ rad, as $\cos(2\pi f_0 t - \pi/2) = \sin 2\pi f_0 t$ and $\sin(2\pi f_0 t - \pi/2) = -\cos 2\pi f_0 t$. This is the case for all frequency components of a real waveform, so we see that the Hilbert transform is equivalent to a wideband (all-frequency) phase shift of $-\pi/2$.

5

Interpolation for Delayed Waveform Time Series

5.1 Introduction

Here we consider the following question: given a time series obtained by regular sampling of some waveform, how do we form the time series of a delayed version of the waveform? Clearly there is no real problem for a delay that is a multiple of the sampling period—instead of the current sample from the undelayed waveform, we just take the correctly delayed sample. The required series could be obtained from a shift register clocked at the sampling rate. Thus, we are left with the problem of generating series corresponding to delays of less than a sampling period. We consider only sampled analytic signals (complex time series), and we show that considerable benefits, in terms of reduced computation, are given if the waveform is sampled at a rate above the minimum required to retain all its information (see Chapter 4)—the case of oversampling.

First, in Section 5.2, we investigate the weights on the taps of a transversal filter required to give the series for the delayed waveform, which are derived without reference to the waveform. This filter is thus suitable for the general case, where any waveform (subject to it being within a given bandwidth) may be taken and where its power spectrum is not necessarily known. We start with the case of the minimum sampling rate and then explore the gains possible with an oversampled waveform. In Section 5.3, we find the weights that give the optimum series in the sense of the least mean square error (or error in power) between the interpolated series and the true series for the delayed waveform. The error arises because to achieve perfect interpolation in principle, ignoring practical problems of finite word lengths and sampling quantization, an infinitely long filter would be required, in general.

Two applications of interpolation are given in Sections 5.4 and 5.5. The first shows a remarkable reduction in computational load in generating simulated radar clutter, which is sampled at a much higher rate (the radar PRF, typically a few kilohertz) than the bandwidth of the clutter waveform (a few tens of hertz). The second shows how interpolation can be used for resampling—generating the sequence of samples that would have been obtained by sampling a signal at a rate different from that actually used.

5.2 Spectrum Independent Interpolation

In this section we show how a finite impulse response (FIR) filter that will achieve the required interpolation can be designed, with the coefficients easily obtained using the rules and pairs method. Generally this requires quite a long filter if the interpolation is to be achieved with high fidelity when sampling at the minimum rate necessary to preserve the full information. More interestingly, we then consider the case where the waveform is sampled at a rate above this minimum—the oversampled case—and find that by taking advantage of this higher rate, very considerable gains, in terms of reducing the filter length and so the required computation, can be achieved for comparable performance.

5.2.1 Minimum Sampling Rate Solution

Given a time series of samples of a continuous waveform, sample values of that waveform at other times can be calculated by taking a weighted combination of the given samples. A suitable set of weights will produce a time series corresponding to samples taken at a certain interval, or delay, after those of the input. This produces a time series corresponding to a delayed version of the waveform. The series itself is not delayed, except perhaps by a number of sample periods; it is otherwise synchronous with the input series. Figure 5.1 illustrates the structure, which is in fact a transversal, or FIR, filter. The delay T between taps is identical with the sampling period, and we note that the output of the center tap, with weight w_0, can be considered to be the undelayed waveform if an overall delay of nT can be accepted. In this

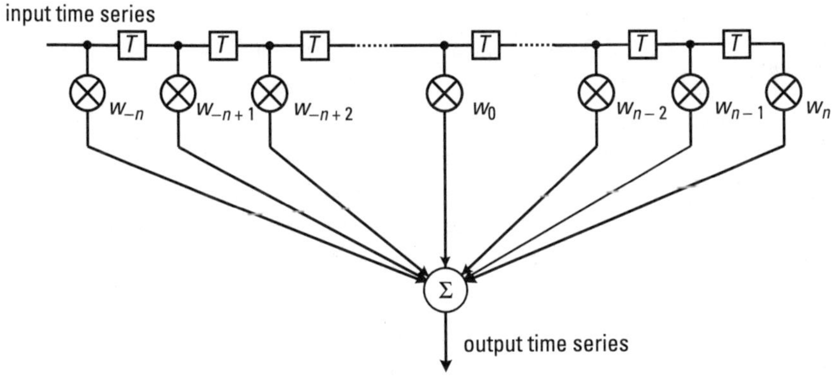

Figure 5.1 FIR filter for interpolation.

case, it is possible to obtain (relatively) negative delays as well as positive ones. (For example, if all the weights were zero except the first, w_{-n}, then the relative delay of the output series would be $-nT$.) We take the time series to be that of a complex baseband waveform of finite bandwidth with spectrum in the band $-F/2$ to $+F/2$, corresponding to an RF or IF waveform of bandwidth F. The minimum sampling rate to retain the information in the waveform is F, and initially we take this to be the sampling rate for the time series, but subsequently we investigate the benefit, from the point of view of more efficient interpolation, of sampling at a higher rate. If the signal waveform is $u(t)$ and the spectrum is $U(f)$, then we can write the identity

$$U(f) = \text{rect}\,(f/F)\,\text{rep}_F\,U(f) \tag{5.1}$$

This states that U is equal to a suitably gated portion of a repetitive form of itself (Figure 5.2).

The (inverse) Fourier transform of this (from P3a, R5, R7a, and R8b) is

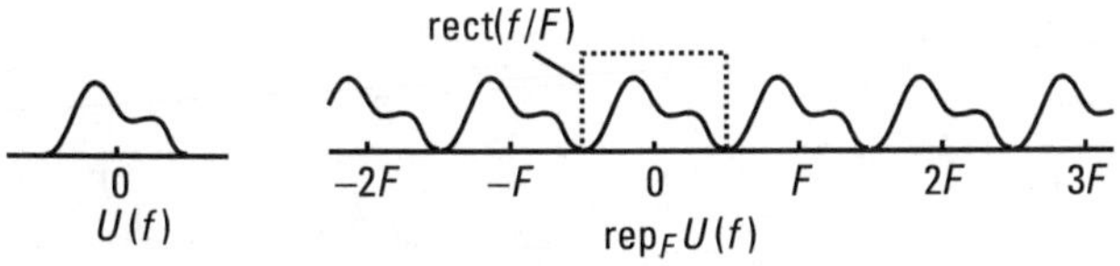

Figure 5.2 Equivalent forms of $U(f)$.

$$u(t) = F \operatorname{sinc}(Ft) \otimes \frac{1}{F} \operatorname{comb}_{1/F} u(t) = \operatorname{sinc}(t/T) \otimes \operatorname{comb}_T u(t) \tag{5.2}$$

where T is the sampling period and $T = 1/F$. The function $\operatorname{comb}_T u(t)$ is a set of δ-functions at intervals T of strengths given by the waveform values at the sampling point [as defined in (2.7)]. Putting the comb function in this form, we have

$$u(t) = F \operatorname{sinc}(t/T) \otimes \sum_{n=-\infty}^{n=\infty} u(nT)\,\delta(t - nT) \tag{5.3}$$

$$= \sum_{n=-\infty}^{n=\infty} u(nT) \operatorname{sinc}\left(\frac{t - nT}{T}\right)$$

This shows how to calculate $u(t)$ at any time t from the given set of samples at times $0, \pm T, \pm 2T, \ldots$, that is, $\{u(nT): n = -\infty < n < +\infty\}$. We place a sinc function, scaled by the sample value, at each sample position and sum these waveforms (Figure 5.3). In particular, if $t = rT$, where r is an integer, then $\operatorname{sinc}[(t - nT)/T] = \operatorname{sinc}(r - n) = \delta_{rn}$, as $\operatorname{sinc}(x) = 0$ for x a nonzero integer and $\operatorname{sinc}(0) = 1$, and we have

$$u(rT) = \sum_{n=-\infty}^{n=\infty} u(nT)\,\delta_{rn} = u(rT)$$

as required (δ_{rn} is the Kronecker-δ).

To determine the function value at time τ (where we only need to consider $|\tau| \leq T/2$), we have, from (5.3)

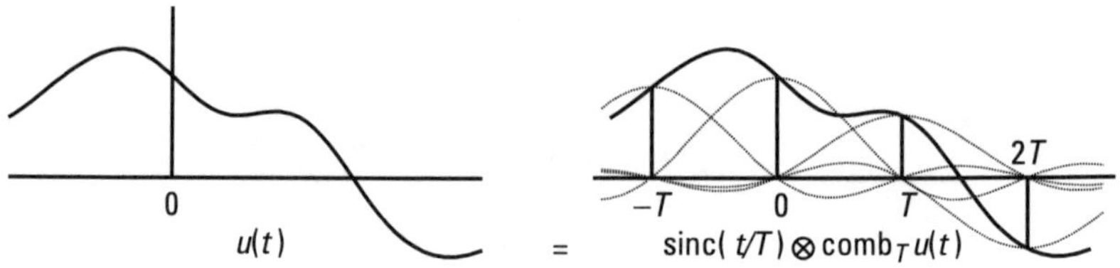

Figure 5.3 Equivalent forms of $u(t)$.

$$
\begin{aligned}
u(\tau) &= \sum u(nT)\,\mathrm{sinc}\left(\frac{\tau - nT}{T}\right) \\
&= u(0)\,\mathrm{sinc}\left(\frac{\tau}{T}\right) + u(T)\,\mathrm{sinc}\left(\frac{\tau - T}{T}\right) + \ldots \\
&\quad + u(-T)\,\mathrm{sinc}\left(\frac{\tau + T}{T}\right) + \ldots \\
&= w_0 u(0) + w_1 u(T) + \ldots + w_{-1} u(T) + \ldots
\end{aligned}
\tag{5.4}
$$

In practice, we cannot obtain $u(\tau)$ exactly, as this requires an infinite number of terms, but the weights applied generally fall (though not necessarily monotonically) for samples further away from the interpolated sample time (within $\pm T/2$ of the center), so we curtail the series when the weights become small. The worst case for interpolation is for delays of $\pm T/2$ at the maximum distance from a sample. The interpolation factor, or weight applied to the output of tap r (that is, to samples at time rT relative to the center tap output), is, in this case,

$$
w_r = \mathrm{sinc}\left(r - \frac{1}{2}\right) = \frac{\sin \pi\left(r - \frac{1}{2}\right)}{\pi\left(r - \frac{1}{2}\right)} = \frac{(-1)^{r+1}}{\pi\left(r - \frac{1}{2}\right)} \tag{5.5}
$$

The tap weights are given in decibel form as the discrete points on the curve in Figure 5.4 for three delay values. For the case of a delay of $0.1T$ (with symbol +), the weight is close to unity for the zero delay tap and falls quite rapidly for the other weights. At a delay of $0.5T$, the weights (given by a dot symbol) are equal for the first two closest taps (numbers 0 and 1) and then fall away rather slowly. For a delay of $0.25T$, the weight pattern (symbol $\times$) is intermediate, but closer to the $0.5T$ case, falling away slowly. If we take -30 dB as the weight level below which we will neglect the contributions, then we see that we need only about 7 taps for the $0.1T$ delay, but 14 at $0.25T$ and 20 at $0.5T$.

5.2.2 Oversampling and the Spectral Gating Condition

For a (complex) waveform of bandwidth F, the minimum sampling rate at which the waveform can be sampled without losing information is F. If we

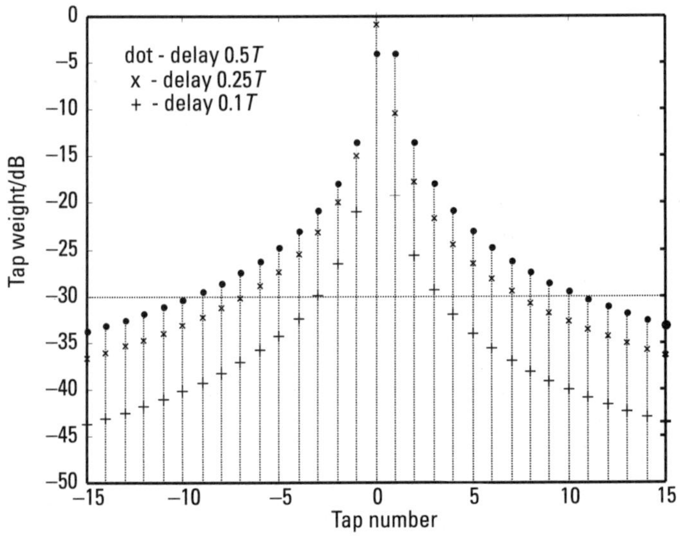

Figure 5.4 FIR filter weights for interpolation at minimum sampling rate.

sample at a lower rate, then the repeating spectra will overlap, and the resulting set of samples would correspond to the result of sampling a slightly different waveform (a distorted form of the waveform) at this lower rate. This effect is known as *aliasing*. However, if we sample at any rate higher than F, no spectral overlapping occurs and we retain all the waveform information, and we could reconstruct the waveform with correct interpolation. This is less efficient than sampling at the minimum rate in the sense that more work is done in sampling than is necessary, but we will see below that it gives the benefit that much more efficient interpolation can be achieved.

Let the sample rate be $F' = qF$, where $q > 1$, so that the interval between samples is now $T' = 1/F' = T/q$. In this case, the spectrum of the sampled waveform repeats at the interval F', which is greater than the width of the basic spectrum, so there are gaps in the spectrum of the sampled waveform, as shown in Figure 5.5. We see that we can put the identity for the waveform spectrum, corresponding to (5.1) above for the minimum sampling rate, in the modified form

$$U(f) = G(f)\,\mathrm{rep}_{F'}\,U(f) \tag{5.6}$$

where an example of the gating function G is shown in Figure 5.5. For (5.6) to be true, we see that there are two conditions that G must satisfy:

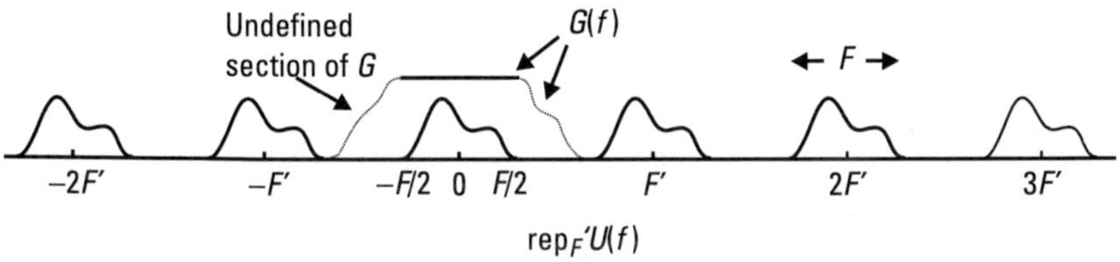

Figure 5.5 Spectrum of time series of *u* sampled at rate *F*′.

$$G(f) = 1 \text{ for } |f| < F/2 \text{ and } G(f) = 0 \text{ for } |f| > F' - F/2 = \left(q - \frac{1}{2}\right)F \tag{5.7}$$

The first of these conditions is to ensure that there is no spectral distortion and the second ensures that there is no aliasing—that is, no energy is included from repeated parts of the spectrum. G is not defined in the regions $[-(q - ½)F, -F/2]$ and $[F/2, (q - ½)F]$ (except that it must remain finite), as there is no spectral power in these regions. Thus we are free to choose G to be of any form as long as it satisfies the conditions in (5.7). In the case of sampling at the minimum rate F, we have $q = 1$, so the regions of free choice are of zero width and we are forced to make G the rect function, as in (5.1). [We could take, as a more general form for the second condition in (5.7), that G should be zero only on all the intervals $|f - nF'| < F/2$ ($n = -\infty$ to $+\infty$)—that is, for all bands of width F centered on all frequencies nF except $n = 0$. However, this will not generally be a useful relaxation of the condition.]

From (5.6), taking the (inverse) Fourier transforms, we have

$$u(t) = (1/F')g(t) \otimes \text{comb}_{T'}\, u(t) = \phi(t) \otimes \text{comb}_{T'}\, u(t) \tag{5.8}$$

where the interpolating function is

$$\phi(t) = (1/F')g(t) = T'g(t) \tag{5.9}$$

and g is the (inverse) Fourier transform of G. Expanding the comb function, we have

$$\begin{aligned} u(t) &= \phi(t) \otimes \sum \delta(t - rT')\, u(rT') \\ &= \sum \phi(t - rT')\, u(rT') \\ &= \sum w_r u(rT') \end{aligned}$$

Thus the weights are given, for a delay τ, by

$$w_r(\tau) = \phi(\tau - rT') \tag{5.10}$$

The samples in this case are at intervals T', so the worst case delay is $T'/2$, smaller than the value at the minimum sampling rate $T/2$; there is, therefore, some easing of the interpolation problem, but this is small compared with that obtainable from good choices of the gating function, as shown below. Before considering these, we take the case of the simplest form of the gating function that takes advantage of oversampling (Figure 5.6). This is $G(f) = \text{rect}\,[f/(2F' - F)]$, or $\text{rect}\,[f/(2q-1)F]$, and the interpolating function is given, from (5.9), by

$$\phi(t) = \frac{(2q-1)F}{qF}\,\text{sinc}\,(2q-1)Ft = \frac{(2q-1)}{q}\,\text{sinc}\,(2q-1)t/T \tag{5.11}$$

The characteristic width of this function, $T/(2q-1)$, is narrower than the sample separation T/q, so we should be able to use fewer taps for a given interpolation performance (such as that given by the −30-dB weight level taken in Section 5.2.1). The weights for a delay $\tau = \rho T'$ are, from (5.10),

$$w_r(\rho) = \frac{2q-1}{q}\,\text{sinc}\,\frac{(2q-1)(r-\rho)}{q} = \frac{\sin(2q-1)X/q}{X} \tag{5.12}$$

where $X = \pi(r - \rho)$. We note that if ρ, and hence τ, is zero, w_r is nonzero for all values of r, unlike the minimum sampling case illustrated in Figure 5.3, where $w_r = 0$, except for w_0, which is 1. Thus, it is not immediately obvious how this sampling method produces the correct values at the sampling points, let alone between them. In particular, $w_0(0) = (2q-1)/q$, which approaches the value 2 for large q. Figure 5.7 illustrates the case where a

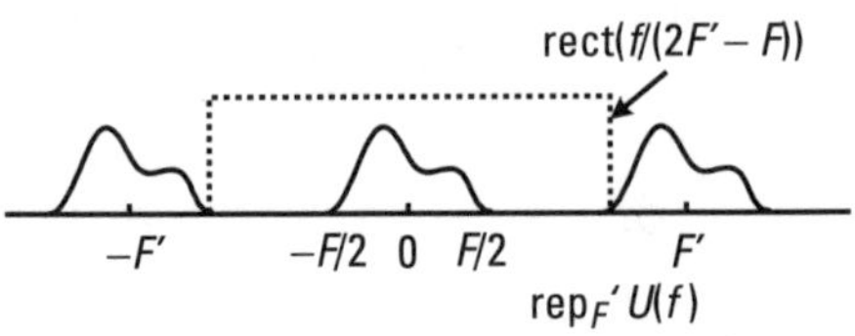

Figure 5.6 Optimum rectangular gate for oversampled time series.

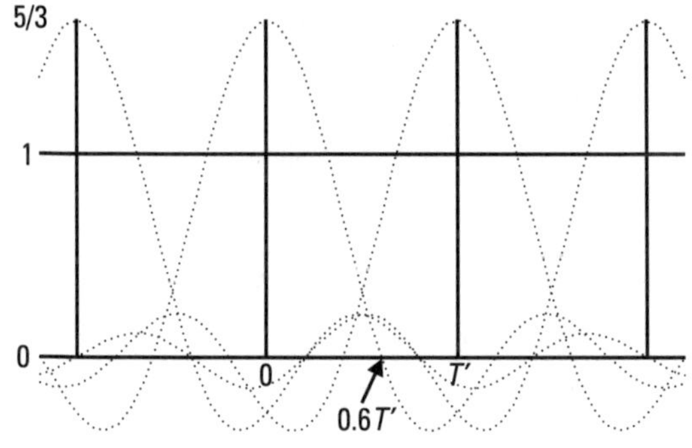

Figure 5.7 Flat waveform oversampled.

flat part of the waveform, with constant value unity, has been sampled at an oversampling rate of $q = 3$. We see that at the sample points the weight value is 5/3, but the contributions from the interpolating sinc functions from nearby sample points are negative, bringing the value down to the correct level.

The weights given by (5.12) for oversampling factors of 2 and 3 are shown in Figure 5.8 for comparison with the values for the minimum sampling rate ($q = 1$) plotted in Figure 5.4. The same set of delays has been taken. These plots show that the weight for the tap nearest the interpolation point (taken to be the center tap here) can be greater than unity, that the weight magnitudes do not necessarily fall monotonically as we move away from this point, and that much the same number of taps is required above a given weight level, such as −30 dB. At first, this last point might seem unexpected—there is no significant benefit from using the wider spectral gate that is possible with oversampling. However, the relatively slow falling off of the tap weight values is a result of the relatively slowly decaying interpolating sinc function, and this in turn is the result of using the rectangular gate with its sharp, discontinuous edges. This is the case whether we have oversampling or not. The solution, if fewer taps are to be required, is to use a smoother spectral gating function, and this is the subject of the next section.

5.2.3 Three Spectral Gates

Trapezoidal

The first example of a spectral gate without the sharp step discontinuity of the rect function is given by a trapezoidal function (Figure 5.9). As illustrated

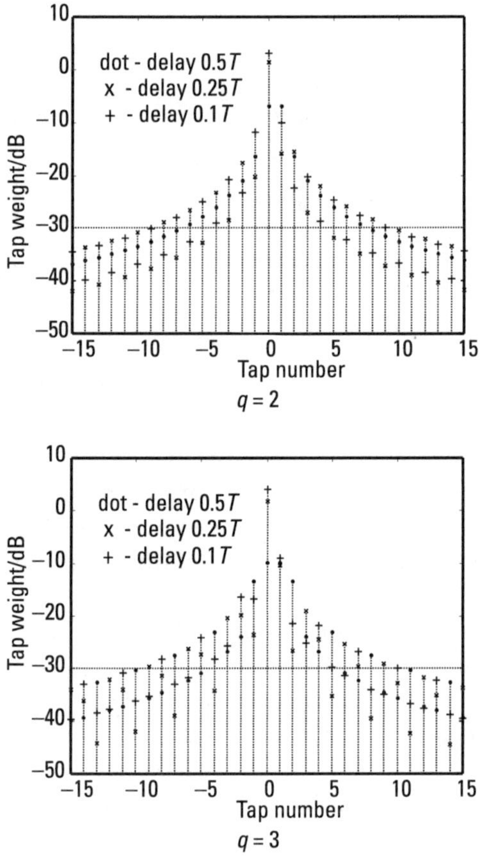

Figure 5.8 FIR interpolation weights with oversampling.

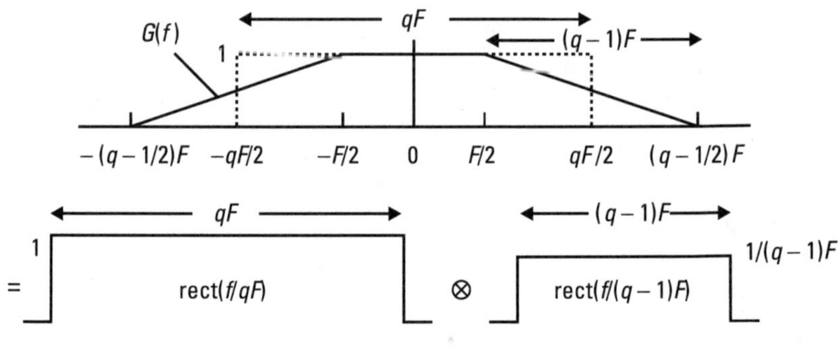

Figure 5.9 Trapezoidal spectral gate.

in this figure and also in Section 3.1, this symmetrical trapezoidal shape is given by the convolution of two rectangular functions with a suitable scaling factor. The convolution has a peak (plateau) level of $(q-1)F$ (the area of the smaller rect function), so we define G by

$$G(f)=\frac{1}{(q-1)F}\operatorname{rect}\left(\frac{f}{qF}\right)\otimes\operatorname{rect}\left(\frac{f}{(q-1)F}\right) \tag{5.13}$$

Thus, on taking the transform, $g(t)=qF\operatorname{sinc}qFt\operatorname{sinc}(q-1)Ft$, and the interpolating function from (5.9) with $T'=1/F'=1/qF$, is

$$\phi(t)=\operatorname{sinc}qFt\operatorname{sinc}(q-1)Ft \tag{5.14}$$

From (5.8) we have

$$u(t)=\operatorname{sinc}qFt\operatorname{sinc}(q-1)Ft\otimes\operatorname{comb}_{1/F'}u(t) \tag{5.15}$$

The interpolating function ϕ is now a product of sinc functions, and this has much lower side lobes than the simple sinc function. To interpolate at time $\tau=\rho T'$, where $0<\rho<1$ (i.e., τ is a fraction of a tap interval), we consider the contribution from time sample r, giving

$$w_r(\tau)=\phi[(r-\rho)T']=\operatorname{sinc}(r-\rho)\operatorname{sinc}[(r-\rho)(q-1)/q] \tag{5.16}$$

Now let $x=r-\rho$ and $y=(q-1)x/q$; then

$$w_r(\tau)=\operatorname{sinc}x\operatorname{sinc}y=\sin X\sin Y/XY \tag{5.17}$$

where $X=\pi x$ and $Y=\pi y$. If we take the case of $\rho=½$, the worst case, as in Section 5.2.1, we have $\sin X=\sin\pi(r-½)=(-1)^{r+1}$, and if we take $q=2$ (sampling at twice the minimum rate), then $\sin Y=\sin\pi(r-½)/2=\pm1/\sqrt{2}$ for r integral. So the magnitudes of the tap weights are

$$\left|w_r\left(\frac{1}{2}\right)\right|=\left|\phi\left(r-\frac{1}{2}\right)\right|=\frac{\sqrt{2}}{\pi^2\left(r-\frac{1}{2}\right)^2} \tag{5.18}$$

Comparing this with (5.5), we see that the weight values now fall very much faster, and this is illustrated in Figure 5.10 for comparison with Figures 5.4 and 5.8.

We see that the number of taps above any given level has been reduced dramatically—above −30 dB, for example, from 20, 15, and 7 at $q = 1$ for the three delays chosen, to 4, 3, and 3 at $q = 2$; and as few as 2, 3, and 2 at $q = 3$. Above the −40-dB level, the number of taps needed at $0.5T$ is found to be 65 at the minimum sampling rate, but only 8 for $q = 2$ and $q = 3$.

Rectangular with Trapezoidal Rounding

The trapezoidal function still has slope discontinuities, though not the step discontinuities that the rect function has. The corners of the trapezoid can be rounded by another rect convolution, to make three convolved rect functions in total. The combination of the two narrower rect functions, the first removing the steps and the second removing the abrupt slope changes, together form a trapezoidal rounding pulse as illustrated in Figure 5.11. As before, the main rect function is of width qF (as in Figure 5.9) and the overall rounding pulse is of base length $(q-1)F$, as this is the space available for the rounding, on each side. Let the two shorter rectangular pulses be of length $\alpha(q-1)F$ and $(1-\alpha)(q-1)F$, where $0 < \alpha \le 0.5$, and then their convolution will be of the required length $(q-1)F$ as shown in the upper part of Figure 5.11. If these pulses are of unit height, then the trapezoidal pulse will be of height $\alpha(q-1)F$, the area of the smaller pulse, so we need to divide by this factor to form a trapezoidal pulse of unit height. The area of the trapezoidal pulse A is the same as that of the wider rectangle, $(1-\alpha)(q-1)F$, and we also have to divide by this factor when we perform the second convolution in order to make the height of G unity, as required. Thus we have

$$G(f) = \frac{\operatorname{rect} f/qF \otimes \operatorname{rect} f/\alpha(q-1)F \otimes \operatorname{rect} f/(1-\alpha)(q-1)F}{\alpha(1-\alpha)(q-1)^2F^2} \tag{5.19}$$

The interpolating function ϕ is given by

$$\phi(t) = (1/qF)g(f) = \operatorname{sinc} qFt \operatorname{sinc} \alpha(q-1)Ft \operatorname{sinc}(1-\alpha)(q-1)Ft \tag{5.20}$$

Let $t = (r-\rho)T'$ as before (with $0 < \rho \le 0.5$), $x = qFt = r - \rho$, and $y = (q-1)x/q$; then

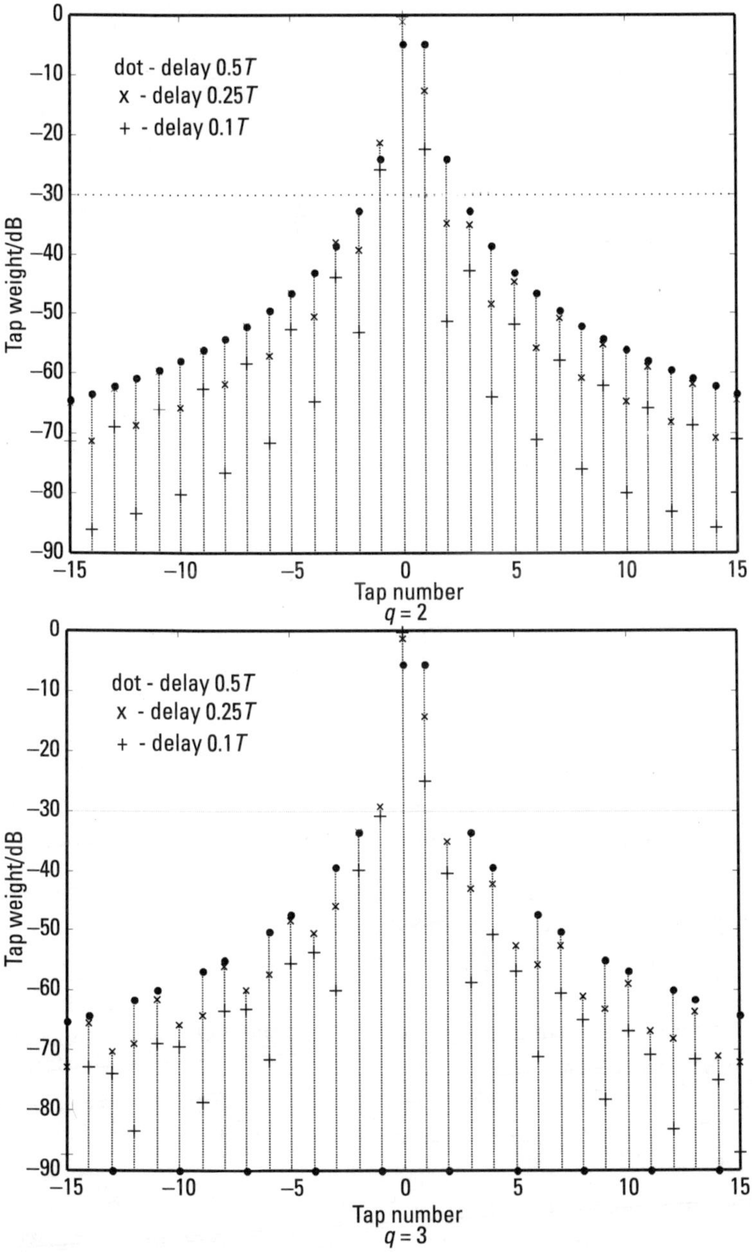

Figure 5.10 Filter weights with oversampling and trapezoidal spectral gate.

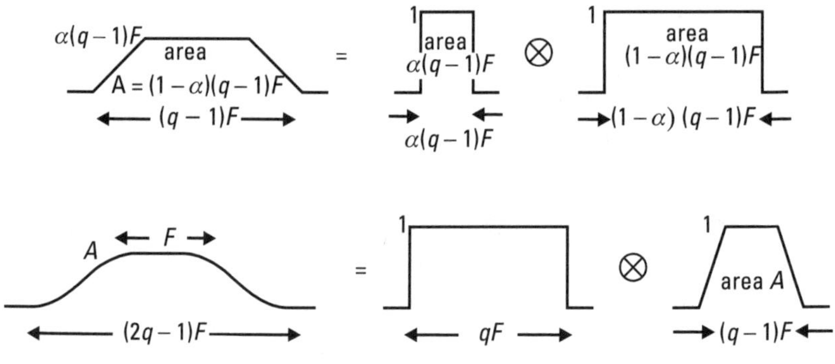

Figure 5.11 Trapezoidal rounding.

$$w_r(\tau) = \phi[(r - \rho)T'] = \text{sinc}\, x \;\text{sinc}\,(1 - \alpha)y \;\text{sinc}\, \alpha y \qquad (5.21)$$

$$= \frac{\sin X \sin Y_1 \sin Y_2}{XY_1Y_2}$$

where $X = \pi x$, $Y_1 = \alpha\pi y$ and $Y_2 = (1 - \alpha)\pi y$.

If $\alpha = 0.5$, we have a triangular pulse for the rounding convolution, but this may make the edge too sharp. As we reduce α, we go through the trapezoidal form towards the rectangular case considered above. The weights for the same three delays as before are plotted in Figure 5.12 for oversampling factors of 2 and 3, for a value of α of 1/3. Again we see that very few taps are needed compared with the rectangular case, and the weight values are seen to be falling away more rapidly than for the simple trapezoidal case, as expected.

Rectangular with Raised Cosine Rounding

Here we use a raised cosine pulse for rounding instead of the trapezoidal pulse above. This pulse is of the form $1 + \cos(af)$, so it has a minimum value of zero and is gated to one cycle width, this being the required value $(q - 1)F$. If $2A$ is its peak value, then the pulse shape (in the frequency domain) is given by $A\,\text{rect}[f/(q - 1)F]\{1 + \cos[2\pi f/(q - 1)F]\}$ (Figure 5.13). This has integral $A(q - 1)F$, due to the raised offset only, as the integral of the single cycle of the cosine function within the rect gate is zero. In order to make the area unity, we take $A = 1/(q - 1)F$. Applying this to the main spectral gating rect function to give the smoothed form, we have

$$G(f) = \text{rect}\,\frac{f}{qF} \otimes \left\{\text{rect}\left[\frac{f}{(q - 1)F}\right] \frac{(1 + \cos[2\pi f/(q - 1)F]}{(q - 1)F}\right\} \qquad (5.22)$$

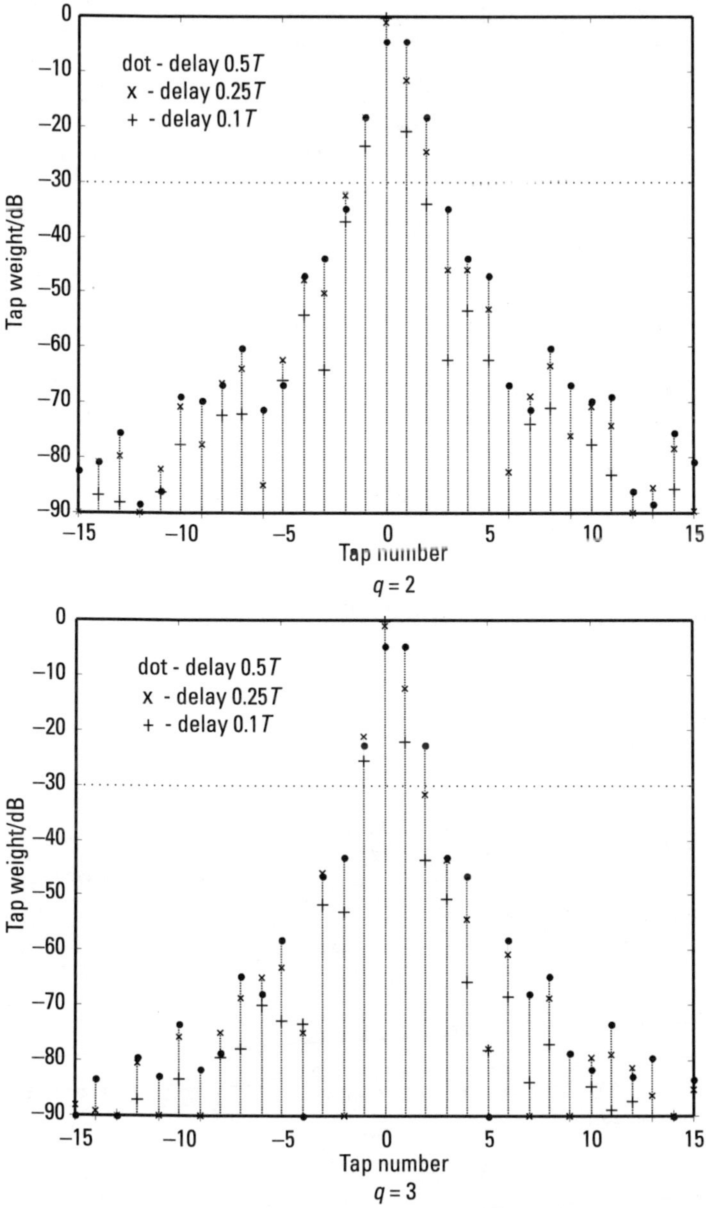

Figure 5.12 Filter weights with oversampling and trapezoidal rounded gate.

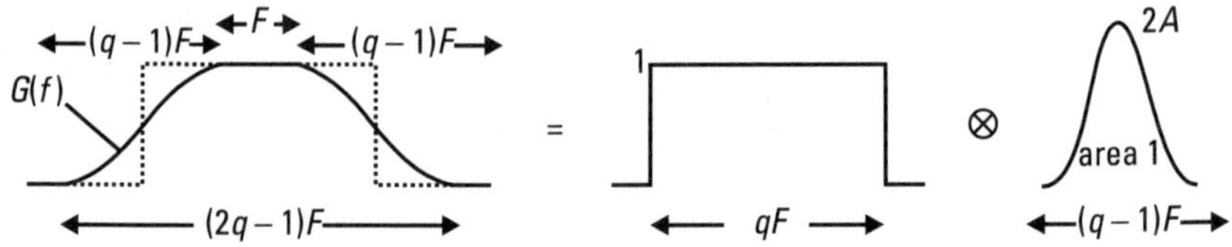

Figure 5.13 Raised cosine rounding.

and

$$g(t) = qF \operatorname{sinc} qFt \left\{ \operatorname{sinc}(q-1)Ft \otimes \left[\delta(t) + \frac{\delta(t-\Delta t) + \delta(t+\Delta t)}{2} \right] \right\}$$

where $\Delta t = 1/(q-1)F$. On performing the δ-function convolutions, the interpolating function is

$$\begin{aligned} \phi(t) &= \frac{1}{qF} g(t) \\ &= \operatorname{sinc} qFt \left\{ \operatorname{sinc}(q-1)Ft + \frac{1}{2}\operatorname{sinc}[(q-1)Ft - 1] + \frac{1}{2}\operatorname{sinc}[(q-1)Ft + 1] \right\} \end{aligned} \tag{5.23}$$

The term in braces {} has much lower side lobes, though a wider main lobe, than the basic sinc function, as should be expected from the form of the gating, or windowing, function G (Hamming weighting). With the same notation as above, we have for the delay $\tau = \rho T'$,

$$\begin{aligned} w_r(\tau) &= \phi[(r-\rho)T'] = g[(r-\rho)T']/qF \\ &= \operatorname{sinc} x \left[\operatorname{sinc} y + \frac{1}{2}\operatorname{sinc}(y-1) + \frac{1}{2}\operatorname{sinc}(y+1) \right] \end{aligned} \tag{5.24}$$

Putting

$$\operatorname{sinc}(y \pm 1) = \frac{\sin \pi(y \pm 1)}{\pi(y \pm 1)} = \frac{-\sin \pi y}{\pi(y \pm 1)}$$

we have

$$w_r(\tau) = \frac{\sin \pi x \sin \pi y}{\pi^2 x} \left[\frac{1}{y} - \frac{1}{2(y-1)} - \frac{1}{2(y+1)} \right] \quad (5.25)$$

$$= \frac{-\sin \pi x \sin \pi y}{\pi^2 xy(y^2 - 1)} = \frac{\sin X \sin Y}{XY(1 - y^2)}$$

Compared with the case of the trapezoidal gate above [see (5.17)], there is an extra factor in the denominator of $1 - y^2$, which is effective in reducing the magnitudes of w_r when r is large. Figure 5.14 shows the weights for the same delays and oversampling factors as before, and we see that the weight values fall even faster than with trapezoidal rounding as a result of the very smooth form of this rounding.

5.2.4 Results and Comparisons

In this section we give the tap weights (in decibels) for the case $\rho = ½$, that is, for the worst-case interpolation, half-way between two taps. For smaller ρ, the weight values will fall faster with r. For small delays (very much less than $T'/2$), oversampling may hardly be needed to keep down the number of taps while maintaining good signal fidelity, but in many applications any delay may be required, and here we evaluate the tap weights for the worst case.

Results for four different interpolation expressions are obtained below, following the different spectral gating functions given above. These are

1. Maximum width rectangular gating (5.12)

$$w_r(\rho) = \sin[(2q-1)X/q]/X \qquad [X = \pi(r - \rho)]$$

2. Trapezoidal spectral gating (5.17)

$$w_r(\rho) = \sin X \sin Y/XY \qquad [Y = (q-1)X/q]$$

3. Gate with trapezoidal rounding (5.21)

$$w_r(\rho) = \text{sinc}\, x \,\text{sinc}\,(1-\alpha)y \,\text{sinc}\, \alpha y = \frac{\sin X \sin Y_1 \sin Y_2}{XY_1Y_2}$$

$$[Y_1 = \alpha Y,\ Y_2 = (1-\alpha)Y]$$

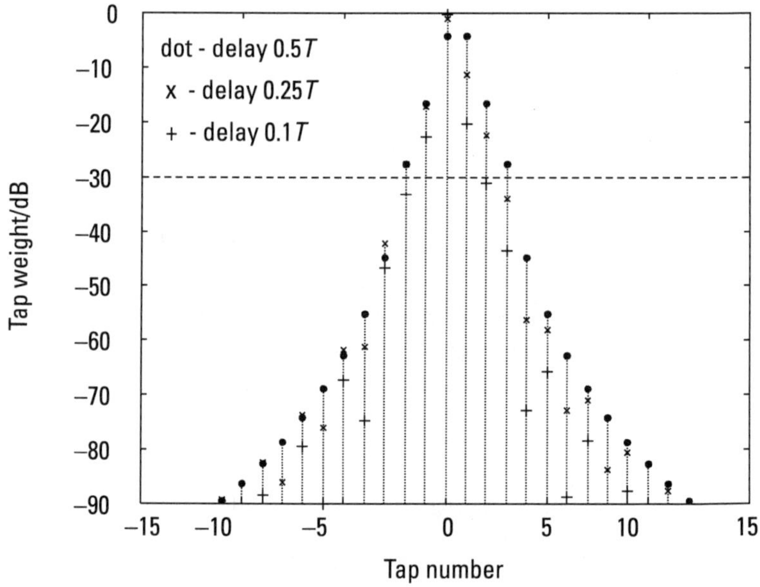

$q = 2$

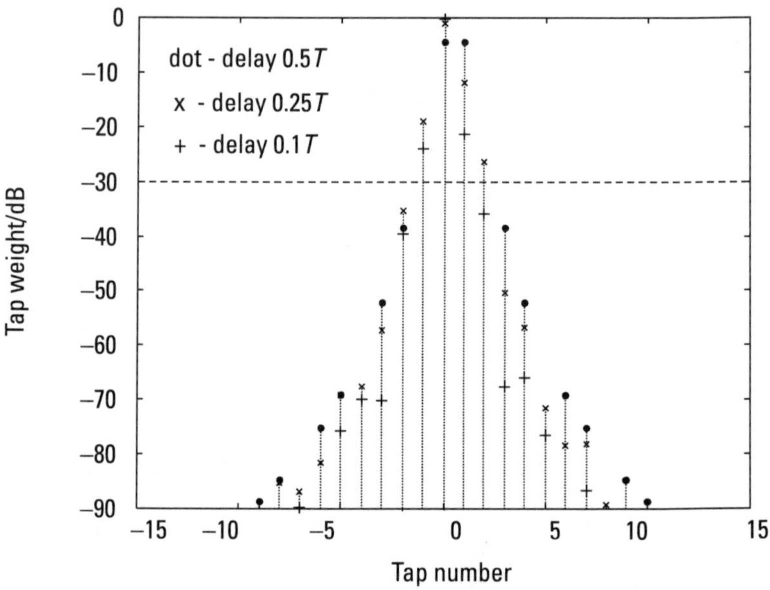

$q = 3$

Figure 5.14 Filter weights with oversampling and raised cosine rounded gate.

4. Gate with raised cosine rounding (5.25)

$$w_r(\rho) = \frac{\sin X \sin Y}{XY(1 - y^2)} \qquad (y = Y/\pi)$$

Figure 5.15 shows, in contour plot form, how the filter tap weights vary with oversampling rate for the worst-case delay of $0.5T$. The tap weights are given in decibel form with tap number along the x-axis and oversampling rate along the y-axis. The contours are at 10-dB intervals. Only integer values for the tap numbers are meaningful, of course, but the expressions above are not restricted to integer values of r, and so contour plots can be drawn. We see how the weight values fall only slowly at $q = 1$, but only a small increase, to 1.2, for example, reduces the levels rapidly. The general pattern is fairly similar for three of the different gating functions, with the trapezoidal rounded gate perhaps the best, but the rectangular gate is very much poorer in rate of fall of coefficient strength with increasing sampling factor. This is consistent with the discussion of Figure 5.8.

5.3 Least Squared Error Interpolation

5.3.1 Method of Minimum Residual Error Power

In Section 5.2 we saw how to approximate the time series for the sampled delayed waveform, given the time series of a sampled waveform. The approximation is not exact, because only a finite set of FIR filter taps can be used in practice. The error in curtailing the filter is not evaluated, because this will depend on the actual waveform, and the approach of that section is independent of the waveform, given that it is of finite bandwidth. In this section a different approach is taken; the question tackled is, given a finite-length filter, what is the set of tap weights that minimizes the error (in power) in the delayed waveform series? To answer this question, we do not need the actual waveform, but only its power spectrum, and some example spectral shapes are taken to illustrate the theory in Section 5.3.2 below.

Figure 5.16 shows the FIR filter model, similar to Figure 5.1, with the waveforms x added. We do not distinguish between the continuous waveforms and the sampled forms, as we know that, correctly interpolated, the sampled series form will give the continuous one exactly for a band-limited signal. If we let the required output waveform be delayed by ρT relative to the waveform $x(t)$ at the center tap, then it is given by

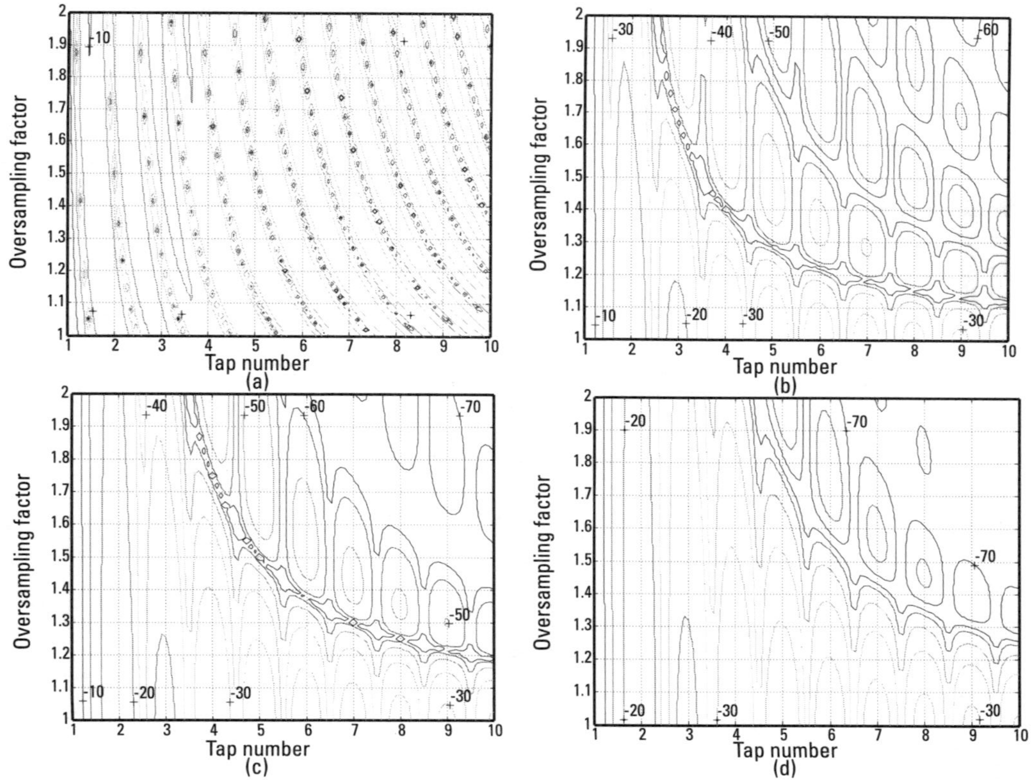

Figure 5.15 Tap weight variation with oversampling rate for four spectral gating functions at delay 0.5*T*: (a) rectangular, (b) trapezoidal, (c) trapezoidal rounded, and (d) raised cosine rounded.

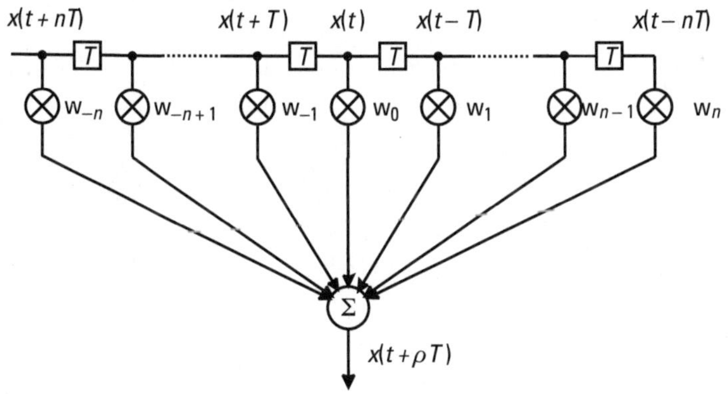

Figure 5.16 FIR filter for interpolation.

$x(t - \rho T)$. T is the sampling period and ρ (where $0 < \rho < 1$) is the delay offset as a fraction of this interval. Although $x(t - \rho T)$ is indicated as the actual filter output in the figure, this could only be achieved with an infinite set of taps, correctly weighted; the actual output, with the tap weights derived below, is a least squared error approximation to this. The error waveform, the difference between the desired output and that given by the FIR filter, is $e(t)$, given by

$$e(t) = x(t - \rho T) - \sum_{k=-n}^{n} x(t - kT) w_k \tag{5.26}$$

Taking the Fourier transform of this equation we have

$$\begin{aligned} E(f) &= X(f) \exp(-2\pi i f \rho T) - \sum_{k=-n}^{n} X(f) \exp(-2\pi i f k T) w_k \\ &= X(f) G(f) \end{aligned}$$

where

$$G(f) = \exp(-2\pi i f \rho T) - \sum_{k=-n}^{n} \exp(-2\pi i f k T) w_k \tag{5.27}$$

The error power p, considered to be a function of the set of weights, is given by

$$p = \int_{-\infty}^{\infty} |E(f)|^2 \, df = \int_{-\infty}^{\infty} |X(f)|^2 \, |G(f)|^2 \, df \tag{5.28}$$

(Here the limits of the second integral could be $-F/2$ and $F/2$, as x is taken to be band-limited, with no spectral power outside this interval.) We suppose that the waveform is of unit power so that $\int_{-\infty}^{\infty} |X(f)|^2 \, df = 1$. From (5.27) we have

$$|G(f)|^2 = 1 - 2 \operatorname{Re}\left[\sum_{k=-\infty}^{\infty} \exp(-2\pi i f \rho T) e^{2\pi i f k T} w_k^* \right] \tag{5.29}$$

$$+ \sum_{h=-\infty}^{\infty} \sum_{k=-\infty}^{\infty} e^{2\pi i f (h-k) T} w_k^* w_h$$

Inserting this into (5.28), we can express the error power in a vector-matrix form by

$$p(\mathbf{w}) = 1 - 2 \operatorname{Re} \{\mathbf{w}^{\mathrm{H}} \mathbf{a}\} + \mathbf{w}^{\mathrm{H}} \mathbf{B} \mathbf{w} \tag{5.30}$$

where we define

$$\mathbf{w} = [w_{-n} w_{-n+1} \ldots w_n]^{\mathrm{T}} \tag{5.31}$$

and the elements of the vector **a** and the matrix **B**, of sizes $2n + 1$ and $(2n + 1) \times (2n + 1)$, respectively, are given by

$$a_k = r[(\rho - k)T] \text{ and } b_{hk} = r[(h - k)T] \tag{5.32}$$

where

$$r(\tau) = \int_{-\infty}^{\infty} |X(f)|^2 \exp(2\pi i f \tau) \, df \tag{5.33}$$

[The raised suffixes T and H indicate matrix transpose and complex conjugate (Hermitian) transpose, respectively]. We see that the components a_k and b_{hk} are values of the autocorrelation function of the waveform x, as r is the

Fourier transform of the power spectrum of x, and this gives the autocorrelation function by the Wiener-Khintchine theorem (Section 2.3).

By differentiating $p(\mathbf{w})$ with respect to $\mathbf{w}^*$ and setting the differential to zero (see Brandwood [1], for example) we find that p is a minimum when the weight vector is $\mathbf{w}_0$ given by

$$\mathbf{w}_0 = \mathbf{B}^{-1}\mathbf{a} \tag{5.34}$$

and the minimum error power is p_0, given by

$$p_0 = 1 - \mathbf{a}^{\mathrm{H}}\mathbf{B}^{-1}\mathbf{a} \tag{5.35}$$

To calculate $\mathbf{w}_0$ and p_0, we only require $\mathbf{a}$ and $\mathbf{B}$, the components of which are all obtained from the autocorrelation function of the waveform. We do not need to postulate particular waveforms for x in order to calculate the optimum weight and the minimum residue, which will depend on the number of taps, the sampling interval, and the delay, but only its spectral power function. Choosing some simple functions, which approximate likely spectra of real signals, it is possible to obtain values for the weights and the residues quite easily. In the next section, we use the rules-and-pairs technique to find the autocorrelation function for six spectral shapes, and in Section 5.3.3 we show some results.

5.3.2 Power Spectra and Autocorrelation Functions

Rectangular Spectrum

In this case we take the power spectrum $|X(f)|^2$ to be given by $(1/F)\,\mathrm{rect}\,(f/F)$, the factor $1/F$ being required to normalize the total power to unity. The Fourier transform of this is $r(\tau) = \mathrm{sinc}\,(F\tau)$, so we have, for the components of $\mathbf{a}$ and $\mathbf{B}$,

$$a_k = \mathrm{sinc}\,[(\rho - k)FT] \text{ and } b_{hk} = \mathrm{sinc}\,[(h - k)FT] \tag{5.36}$$

The minimum sampling rate is equal to the bandwidth F, so the sampling period is $T = 1/F$, but more generally, if the sampling rate is qF, then we have $T = 1/qF$ or $FT = 1/q$, so that (5.36) becomes

$$a_k = \mathrm{sinc}\,[(\rho - k)/q] \text{ and } b_{hk} = \mathrm{sinc}\,[(h - k)/q] \tag{5.37}$$

Triangular Spectrum

We can form a triangular shape of base width F as the convolution of two rectangular functions of width $F/2$, as in Section 3.2. This has a peak value of $F/2$. In order to have a total area of unity, representing the total power in the power spectrum, we require a peak value of $2/F$, so the spectrum and the autocorrelation function are given by

$$|X(f)|^2 = (4/F^2)\,\mathrm{rect}\,(2f/F) \otimes \mathrm{rect}\,(2f/F) \text{ and } r(\tau) = \mathrm{sinc}^2\,(F\tau/2) \tag{5.38}$$

The required coefficients are thus

$$a_k = \mathrm{sinc}^2\,[(p-k)/2q] \text{ and } b_{hk} = \mathrm{sinc}^2\,[(h-k)/2q] \tag{5.39}$$

Raised Cosine Spectrum

The raised cosine power spectrum of unit area is given by $(1/F)[1 + \cos(2\pi f/F)]\,\mathrm{rect}\,(f/F)$. The transform of the raised cosine, as in Section 3.4, gives the autocorrelation function $\mathrm{sinc}\,(F\tau) + ½\,[\mathrm{sinc}\,(F\tau - 1) + \mathrm{sinc}\,(F\tau + 1)]$, and hence

$$a_k = \mathrm{sinc}\,[(p-k)/q] + \frac{1}{2}\{\mathrm{sinc}\,[(p-k)/q - 1] + \mathrm{sinc}\,[(p-k)/q + 1]\} \tag{5.40a}$$

and

$$b_{hk} = \mathrm{sinc}\,[(h-k)/q] + \frac{1}{2}\{\mathrm{sinc}\,[(h-k)/q - 1] + \mathrm{sinc}\,[(h-k)/q + 1]\} \tag{5.40b}$$

Gaussian Spectrum

The region of the domain over which the Gaussian, or normal, distribution function is nonzero (its support) is unbounded, so there is, strictly, no minimum sampling (or Nyquist) frequency F corresponding to sampling that will represent this function exactly. However, we can approximate the spectrum, for practical purposes, by taking F to be the bandwidth at which the spectral power density has fallen to some low level, A decibels below the spectral peak, such that sampling at frequency F produces an acceptable low level of aliasing. This defines the variance of the spectrum as $\sigma^2 = F^2/1.84A$. The normalized spectrum is

$$|X(f)|^2 = \frac{1}{\sqrt{2\pi\sigma}} \exp(-f^2/2\sigma^2) \tag{5.41}$$

and its transform, from P5 with R5, is

$$r(\tau) = \exp(-2\pi^2\sigma^2\tau^2) \tag{5.42}$$

Expressing the variance in terms of the spectral limit level, A, we obtain

$$a_k = \exp[-2\pi^2\sigma^2(\rho - k)^2/1.84A^2q^2] \text{ and} \tag{5.43}$$
$$b_{hk} = \exp[-2\pi^2\sigma^2(h - k)^2/1.84A^2q^2]$$

Trapezoidal Spectrum

As in Section 3.1, we form a symmetrical trapezium with a base of width F and a top of width aF $(0 < a < 1)$ by the convolution of two rect functions of width $(1 - a)F/2$ and $(1 + a)F/2$, these being the widths of the sloping edges and the half-height width, respectively (as in Figure 3.1). Using unit rect functions gives a peak height of $(1 - a)F/2$, which would give an area of $(1 - a)(1 + a)F^2/4$, so we have to divide by this factor to give the normalized spectrum:

$$|X(f)|^2 = [4/(1 + a)(1 - a)F^2]\,\text{rect}[2f/(1 - a)F] \otimes \text{rect}[2f/(1 + a)F] \tag{5.44}$$

The transform is

$$r(\tau) = \text{sinc}[(1 - a)F\tau/2]\,\text{sinc}[(1 + a)F\tau/2] \tag{5.45}$$

as shown in Figure 3.2, with $a = 1/3$. We note that taking $a = 0$ or $a = 1$ gives the results for the rectangular and triangular spectral cases, respectively, as limiting cases of the trapezoidal form. Finally, we have

$$a_k = \text{sinc}[(1 - a)(\rho - k)/2q]\,\text{sinc}[(1 + a)(\rho - k)/2q] \tag{5.46a}$$

and

$$b_{hk} = \text{sinc}[(1 - a)(k - h)/2q]\,\text{sinc}[(1 + a)(k - h)/2q] \tag{5.46b}$$

5.3.3 Error Power Levels

The error power, given in (5.35), is given in contour plot form in Figure 5.17 for two of these spectral shapes, the rectangular, using (5.37), and the raised cosine, using (5.40). These give the powers as a function of both the number of taps used and the oversampling factor. Although the contour lines, which are at 5-dB intervals, are continuous, only the values at integral tap numbers are meaningful, of course. These plots show that even modest oversampling rates of 20% or 30% are effective in reducing the number of taps for a given required mismatch level or, alternatively, greatly reducing the mismatch power for a fixed number of taps. The general patterns for these two spectra are quite similar, though the more compact raised cosine spectrum has lower mismatch power than the rectangular spectrum at the same parameter values, as might be expected. The results for the other spectral shapes are similar, and are generally between these two.

Figure 5.18 presents results for the rectangular spectrum with an expanded range of taps and a reduced range of oversampling factor. We see that even with 100 taps, the mismatch power when sampling at the minimum rate is over −25 dB, while with 10% oversampling this level is achieved using only 10 taps, and at 25% only 5 taps are required. We also see that, using 20 taps, the mismatch power at the minimum sampling rate is about −12 dB, but this falls to −50 dB at an oversampling rate of only 1.15. These figures show that even with quite low oversampling rates, considerable reductions in computation for a given performance level or considerable improvement in performance for given computational effort is achievable.

5.4 Application to Generation of Simulated Gaussian Clutter

Here we take a particular example showing that taking advantage of oversampling can give a very substantial saving in computation. The problem considered is to generate simulated clutter, as seen in a given range gate, for modeling radar performance. In this case, the clutter is taken to have a complex amplitude distribution, which is normal (or Gaussian), and also has a Gaussian power spectrum. We show first, in Section 5.4.1, that the required waveform can be generated by an FIR filter fed with a sequence of pseudorandom samples from a normal distribution at the required sample rate, which is the radar pulse repetition frequency (PRF). As the bandwidth of the clutter waveform is very much lower than the radar PRF, the clutter

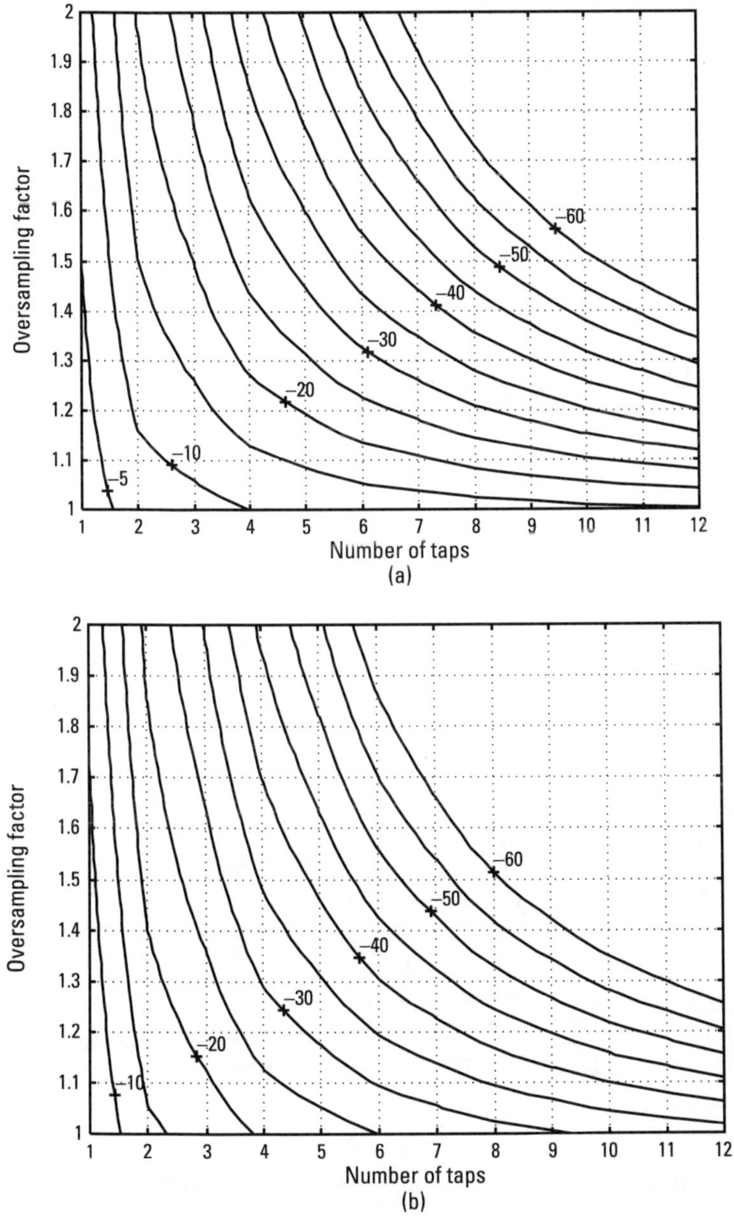

Figure 5.17 Mismatch powers for two power spectra: (a) rectangular spectrum, and (b) raised cosine spectrum.

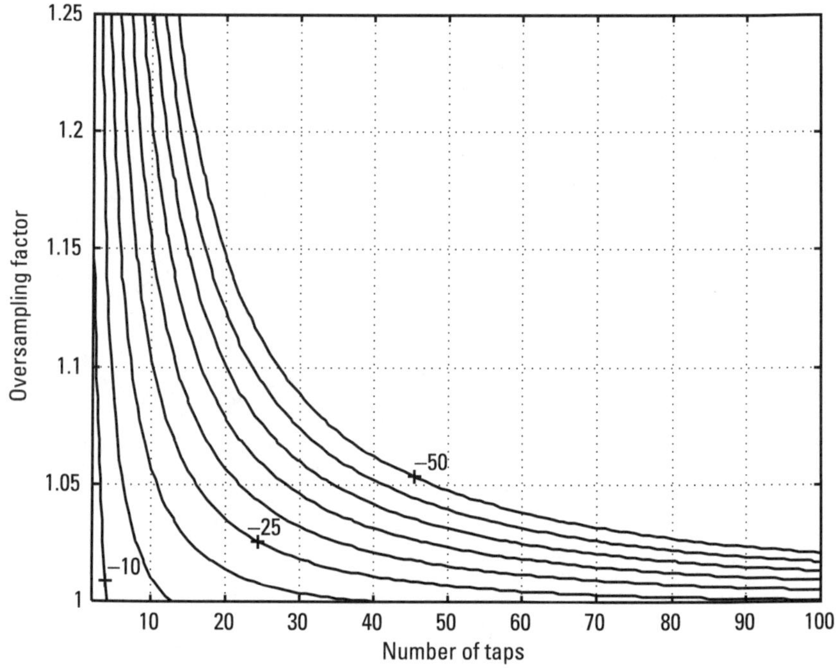

Figure 5.18 Mismatch power for rectangular spectrum.

waveform is greatly oversampled and the cost in computation is high. (Despite high speeds of computation, large, complex simulations, perhaps requiring clutter in many range gates, as in this radar example, can take significant times to carry out, and efficient computation is of value.) In Section 5.4.2, we show that the clutter waveform can be generated at a much lower sampling rate, though still oversampled, and then efficient interpolation is used to give the samples at the PRF, as required. This is shown to reduce the overall computation requirement by a very large factor. The parameters we use for this example are 10 kHz for the PRF F, and 10 Hz root-mean-square (rms) for the spectrum of the clutter waveform.

5.4.1 Direct Generation of Gaussian Clutter Waveform

Any linear combination of independent normally distributed sequences will also be normally distributed (see Mardia et al. [2], for example). An FIR filter of length L fed with a sequence of samples from a normal distribution forms a linear combination of L samples and will produce output samples at intervals L that are independent and normally distributed. The output samples that are at less than L sample intervals apart are not independent, as they are linear combinations of partially overlapping sets of samples, and the choice of FIR filter weights will determine the degree of dependence between successive samples—that is, the rate of change of the values or, equivalently, the frequency spectrum of the output sequence. If $|H(f)|^2$ is the power spectrum required, then the square root of this gives (within an arbitrary phase factor) the required amplitude spectrum and the (inverse) Fourier transform of this gives the required filter impulse response. In the case of an FIR filter, the filter weights, or tap coefficients, are set to the sampled values of the required impulse response (Figure 5.19). (It is clear that an impulse at the input will emerge at the output as a series of impulses scaled by the coefficients, and this is the filter impulse response.) If F is the bandwidth of the spectrum, then we know, from Chapter 4, that the sampling rate must be F or greater, and so the delays between taps in the FIR filter must be $1/F$ or less. The intervals at which the Gaussian impulses are generated and fed into the filter to give the output Gaussian sequence must match this delay, of course.

In the case of a Gaussian spectrum of standard deviation σ and 3-dB bandwidth 2.36σ, we have

$$H(f)^2 \sim \exp(-f^2/2\sigma^2) \tag{5.47}$$

(where we write ~ to indicate that we are not concerned here with the particular scaling factor). As in Section 5.3.2 above, we have an infinite frequency region over which the spectral power density is finite, so we make the approximation of limiting the spectrum to a bandwidth $2r\sigma$ such that the power density at $\pm k\sigma$ is small enough to allow us to neglect the spectral tails and hence the aliasing power. At these points we have $H(\pm r\sigma)^2 = \exp(-r^2/2)$, and this has fallen to −35 dB, as a suitable low level below the peak, at $k = \sqrt{(7 \ln 10)} \approx 4$. In this case, the total bandwidth, outside which there is considered to be negligible spectral power, is 8σ. In this example, this is 80 Hz, which is very low compared with the PRF of 10 kHz, and we see that the required clutter waveform is oversampled by a factor of 125.

From (5.47) we have

$$H(f) \sim \exp(-f^2/4\sigma^2) = \exp\left[-\pi\left(f/2\sigma\sqrt{\pi}\right)^2\right]$$

and so, using P5 and R5, we obtain, for the required filter impulse response,

$$h(t) \sim \exp\left[-\pi\left(2\sigma\sqrt{\pi}t\right)^2\right] = \exp(-4\pi^2\sigma^2t^2) \tag{5.48}$$

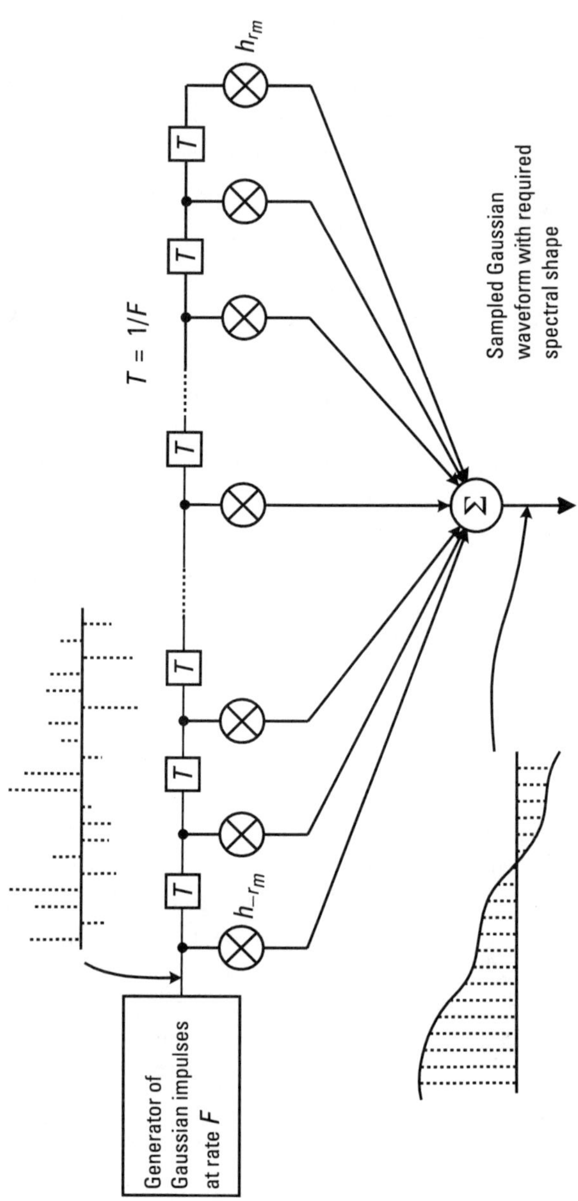

Figure 5.19 FIR filter for Gaussian waveform generation.

The FIR filter coefficients from the sampled impulse response are given by

$$h_r = h(rT) = \exp(-4\pi^2\sigma^2 r^2 T^2) \tag{5.49}$$

where $T = 1/F$ is the sampling interval. If we take coefficients to the −40-dB level, then we have $8\pi^2\sigma^2 r_m^2 T^2 = 4\ln(10)$, or

$$r_m = \frac{\sqrt{\ln(10)/2}}{\pi}\frac{F}{\sigma} = 0.342\frac{F}{\sigma} \tag{5.50}$$

where $\pm r_m$ are the indexes of the first and last coefficients.

We can now estimate the amount of computation required to produce the simulated clutter directly. With $F = 10^4$ Hz and $\sigma = 10$ Hz, we see that $r_m = 342$, so there are 685 taps, and this is the number of complex multiplications needed for each output sample (in addition to generating the inputs from a normal distribution).

5.4.2 Efficient Clutter Waveform Generation Using Interpolation

In this case we generate Gaussian clutter with the required bandwidth but at a much lower sampling rate f_s, and then interpolate to obtain the samples at the required rate F (Figure 5.20). Thus we will need F/f_s times as many interpolations as samples. From Section 5.2 above we know that with moderate oversampling rates, we can achieve good interpolation with very few taps.

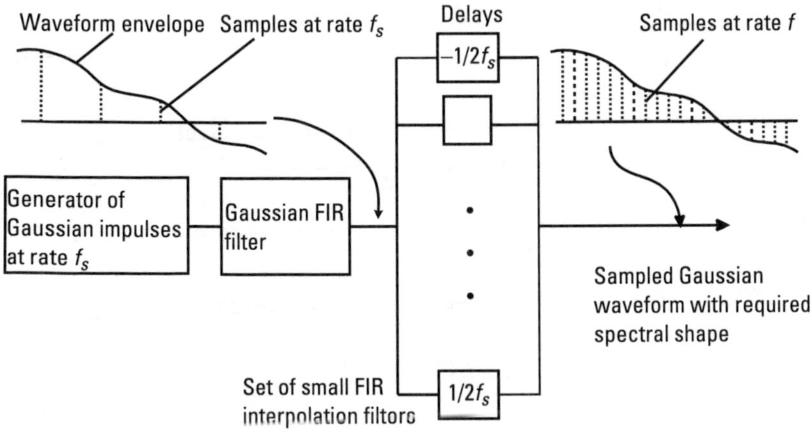

Figure 5.20 Gaussian waveform generation with interpolation.

Let the number of taps in the interpolation filter be m and the number in the Gaussian FIR filter is, from (5.50), $0.684f_s/\sigma$ (+1, which we neglect), so that the average number of complex multiplications per output sample is

$$\nu = m + (0.684f_s/\sigma)/(F/f_s) = m + 0.684f_s^2/\sigma F \qquad (5.51)$$

In Figure 5.12 we see that with an oversampling factor of 3, we need only four taps, weighted above the –40-dB level, to interpolate up to the maximum time shift of half the sampling interval. Using these figures, we have $m = 4$ and $f_s = 24\sigma$ (as the effective bandwidth of the waveform is taken to be 8σ in Section 5.3.1 above), and from (5.51) we obtain $\nu = 4.4$, a factor of over 150 lower than in the direct sampling case. There will have to be $F/2f_s$ sets of four weights (or 21 sets in this example) to interpolate from $-1/2f_s$ to $+1/2f_s$.

5.5 Resampling

An application of interpolation is to obtain a resampled time series. In this case, data has been obtained by sampling some waveform at one frequency F_1, but the series that would have been obtained by sampling this waveform at a different frequency F_2 is now required. We consider first the case where F_1/F_2 is rational and so can be expressed in the form n_1/n_2, with n_1 and n_2 mutually prime (with no common factor). To illustrate the method, we take $n_1 = 4$ and $n_2 = 7$, as shown in Figure 5.21. Over a time interval $T = n_1T_1 = n_2T_2$, the pattern repeats, where $T_1 = 1/F_1$ and $T_2 = 1/F_2$, and if the output sequence is timed so that some samples are at zero shift relative to the input, then there will be further time shifts of $\pm\Delta T$, $\pm 2\Delta T$, . . . , up to $\pm(n_2 - 1)/2$ for n_2 odd, or $-n_2/2 + 1$ and $+n_2/2$ for n_2 even, where $\Delta T = T/n_1n_2$. In Figure 5.21 the required time shifts for the different

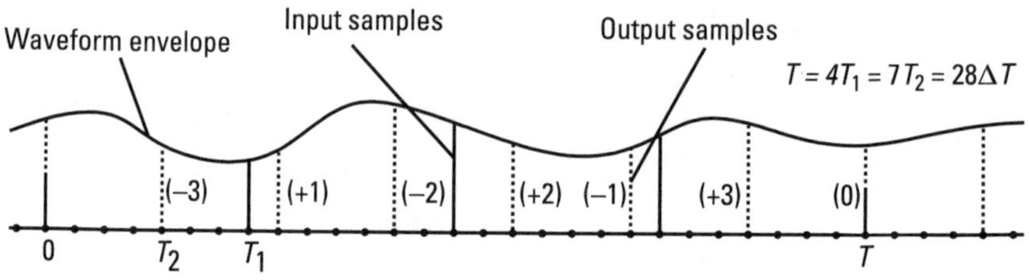

Figure 5.21 Resampling.

pulses are shown in units of ΔT, and we see that the values required are from $-3\Delta T$ to $+3\Delta T$. Over a period of four input pulse intervals, there are seven output pulses, as required, with seven different delays, one being zero. We also see that if the frequency ratio were inverted in this figure, so that the input samples are shown by the dashed lines and the outputs by the continuous lines, then time shifts of −1, +2, +1, and 0 only, relative to the nearest input sample, are required.

If the input sequence is oversampled, we can use the results of Section 5.3.2 above to reduce the size of the sampling FIR filters and so achieve quite economical resampling, requiring only a few multiplications for each output sample. Only $n_2 - 1$ time shifts are needed, and the number of distinct vectors defining the FIR filter coefficients is only $(n_2 - 1)/2$ (n_2 odd) or $n_2/2$ (n_2 even) (as the set of coefficients is the same for positive and negative shifts, applied in reverse order, with a shift of the input sequence) and these can be precomputed and stored. If the output sequence is at a rather higher frequency than the input, as in Figure 5.21, the maximum time shifts, up to half an output sample interval, will be rather less than half an input interval, and this can also be used to reduce the length of the FIR interpolation filters, as shown in the figures in Section 5.2. The processing need not be in real time, of course, with the input and output pulses arriving and departing at the actual intervals specified. The input data could be stored after sampling in real time, of course, and the output sequence could then be generated at leisure, these samples being the values that would have been obtained by real-time sampling at the new frequency. However, if real-time resampling is required, for example on continuous data, then economical computation, as outlined above, could be particularly useful.

If the frequency ratio is not rational, some modifications are necessary. In the case of a block of stored data, it may be acceptable to find a good rational approximation to this ratio. As this is an approximation, the output frequency will not be exactly the specified frequency, and if the waveform is regenerated as if the samples were at this frequency (for example, by a standard sound card in the case of audio data), then there will be a slight frequency scaling of the whole signal. In the case of continuous, real-time data, this would require dropping or inserting a sample from time to time, generally causing an unacceptable distortion of the sound. An alternative would be to calculate accurately the required delay and then the FIR filter tap weights, using equations from Section 5.2. Alternatively, the calculated delay could approximate the nearest of a suitably fine set of values over the half output sample period (positive or negative), and the precalculated set of weights for this delay would be applied.

5.6 Summary

In this chapter we have shown how the rules-and-pairs method can be used to obtain simply results in the field of interpolation for sampled time series, providing insight into the underlying principles. The first main application was to find the FIR filter weights that would provide interpolation for any band-limited signal. In principle, this filter will be infinitely long for perfect interpolation, so in practice a finite filter will always give only an approximation to the correct interpolated waveform. However, a filter of suitable length will give as good an approximation as may be required. For waveforms sampled at the minimum rate, this could be quite long (perhaps 100 or more taps for good fidelity), but if the sampling is at a higher rate (i.e., the waveform is oversampled), the filter length for a given performance is found to fall quite dramatically. This saving in computation could be valuable in large simulations or in providing real-time-delayed waveforms in wide-bandwidth systems, for example.

This first approach does not give a definite estimate of the accuracy of the interpolated waveform, which could be measured, for example, by comparing this waveform, from the FIR filter, with the exact delayed waveform. This will depend on the spectrum of the waveform, and no particular spectrum, within the specified finite bandwidth, is assumed. This is the subject of the second approach, which is to define the filter that will minimize the power in the error signal, the difference between the interpolated series and the exact series, for a given power spectrum. In this case, a few simple spectral shapes were taken to illustrate the technique. In practice, the actual signal spectrum could perhaps be considered a good approximation to one of these. Again, oversampling can be used to reduce greatly the filter length and the number of multiplications for each output sample.

Two applications of interpolation were studied. The first was for the case of generating a greatly oversampled Gaussian waveform. It was shown that generating the Gaussian waveform at a much lower oversampled rate and then interpolating could give a very great reduction (two orders of magnitude) in the amount of computation needed. The second example was the case of resampling, where a sample sequence is required corresponding to having sampled a waveform at a different rate from that actually used. (The previous example is a special case of resampling, where the output frequency is a simple multiple of the input.) Again, this process could be made considerably more economical if the input sequence is oversampled. These examples may not solve any reader's particular problem, but they may provide indications of how to do so, in particular with the simplification and clarity given by the rules and pairs approach.

References

[1] Brandwood, D. H., "A Complex Gradient Operator and Its Application in Adaptive Array Theory," *IEE Proc.*, Vol. 133, Parts F and H, 1983, pp. 11–16.

[2] Mardia, K. V., J. T. Kent, and J. M. Bibby, *Multivariate Analysis*, New York: Academic Press, 1979.

6

Equalization

6.1 Introduction

In this chapter we consider the problem of compensating for some known frequency distortion over a given band. One form of distortion is an unwanted delay, and the resulting distortion is a phase variation that is linear with frequency. This particular case, of delay mismatch, was the subject of Chapter 5, and the method of correction, or equalization, used in Section 6.5 is basically the same as that in Section 5.3. However, we are also concerned with other forms of frequency distortion, and in this chapter the approach is more general, and amplitude variation over the band is included also. In order to do this, a new Fourier transform pair is introduced in Section 6.3, the ramp function, which is a linear slope across the band, and its transform, the snc_1 function, which is the first derivative of the sinc function. In fact, a set of transform pairs is defined that are the integer powers of the frequency across the band (ramp^r) and the derivatives of the corresponding order of the sinc function (snc_r). The sinc and rect functions are seen to be the first (or zeroth order) members of these sets. With these results, any amplitude variation, expressed as a polynomial function of frequency across the band of interest, has a Fourier transform that is a sum of snc_r functions. A simple example of amplitude equalization is given in Section 6.4.

The method of equalization outlined in Section 6.2 is based on minimizing a weighted mean squared error across the band. The error at each frequency is the (complex) amplitude mismatch between the equalized result (normally imperfect) and the ideal, or perfectly equalized, response. The weighting, as in Section 5.3, is given by the spectral power density function of the signal. This has the advantage that the equalization will tend to be best where there is most signal power, and hence the effect of mismatch would be the most serious. If no weighting is required (for example, if the signal spectrum is totally unknown and uniform emphasis across the band is considered most appropriate), then we simply replace the spectral function with the rect function. It is not likely that the spectrum need be accurately known and specified in practice, as a reasonable approximation to the spectral shape will give a result close to that given by an exact form and considerably better than the rather unrealistic unweighted (or constant) shape defined by the rect function, which gives full weight up to the very edges of the band, where normally the signal power will have fallen to a negligible level. Thus, as in Section 5.3, simplifying the spectrum to one of a few tractable forms should be satisfactory. Suitable forms to choose from include the normal (or Gaussian) shape, the raised cosine, or the (symmetric) trapezoidal shape.

In Sections 6.6 and 6.7, we apply the theory given in Sections 6.2 and 6.3 to a specific problem, that of forming broadband sum and difference beams as required for radars using monopulse. A simple example is taken for the array to be used of a 16-element regular linear array to illustrate the application. It would not be difficult to extend the problem to larger, perhaps planar (two-dimensional), arrays—this would increase the number of channels to be equalized, each with its own compensation requirement, but the actual form of the equalization calculation is essentially the same in each case, with different parameters. Thus, although this simple array may not be particularly likely to be used in practice, it is quite adequate to illustrate the benefit of equalization in this application, showing a striking improvement with quite modest computational requirements, given a moderate degree of oversampling.

The radar sum beam (i.e., its normal search beam, giving maximum signal to noise ratio) only requires delay compensation, and this could be provided for each element by the results of Section 5.3. However, Section 6.6 includes results for the full array response with equalization, not considered in Chapter 5, and also provides an introduction to Section 6.7, where the difference beam is considered. This beam, which can be defined as a derivative (in angle) of the sum beam, is used for fine angular position measurement. For this example we carry out equalization in each channel in amplitude as well as phase, and the results of Section 6.3 are now required.

6.2 Basic Approach

The problem to be tackled is that of compensating for a given frequency-dependent distortion in a communications channel, as illustrated in Figure

6.1. A waveform u with baseband spectrum U is received with some channel distortion G such that at (baseband) frequency f, the spectral component received is $G(f)U(f)$ instead of just $U(f)$. The signal is then passed through a filter with frequency response $K(f)$ such that the output spectrum $K(f)G(f)U(f)$ is close to the undistorted signal spectrum $U(f)$. Clearly, the ideal required filter response at frequency f is simply $K(f) = 1/G(f)$, but in practice this filter may not be exactly realizable, for example, if it is an FIR digital filter (except in the unlikely case that K consists of a set of δ-functions corresponding to a number of delays at multiples of the sampling frequency). In this case, we design the filter to give a best fit, in some sense, of $K(f)G(f)U(f)$ to $U(f)$ over the signal bandwidth. In fact, the fit we choose is the least squared error solution, a natural and widely used criterion, which has the advantage of yielding a tractable solution, at least in principle, and this is found to require the application of Fourier transforms. In order to compensate for G, we need to know the form of this function. This may be known from the nature of the system, as in the application in Sections 6.6 and 6.7 below, or a reasonable estimate may be available from channel measurements. In Figure 6.1 we show the incoming signal on a carrier, at frequency f_0, which is generally the case for radio and radar waveforms, and this is down-converted to complex baseband (often in more than one mixing process) and, we assume, digitized for processing, including equalization and detection.

The amplitude error between the filter output and the desired response in an infinitesimal band δf at frequency f is given by $[K(f)G(f)U(f) - U(f)]\,\delta f$, so the total squared error is

$$\int_{-\infty}^{\infty} |K(f)G(f) - 1|^2\,|U(f)|^2\,df \tag{6.1}$$

We note that as the signal spectrum U is included in the error expression, we will actually perform a *weighted* squared error match of KG to unity at all frequencies (the *equalized* solution), where the weighting function is the

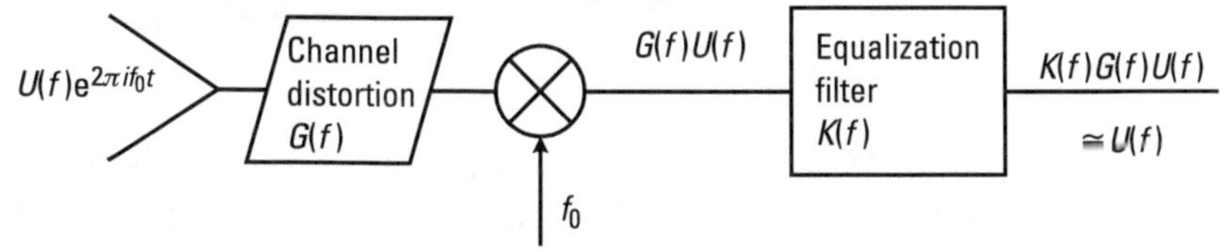

Figure 6.1 Equalization in a communications channel.

spectral power density function of the signal. This means that more emphasis is placed on compensating for distortion in regions where there is more signal power, which is generally preferable to compensating with uniform emphasis over the whole band, including parts where there may be little or no signal power.

The equalizing filter is of the form given in Figure 5.1 or Figure 5.16, and if the filter coefficients are given by v_r for delay rT, where T is the sampling period, then the impulse response of the filter, of length $2n + 1$ taps, is

$$k(t) = \sum_{r=-n}^{n} v_r \delta(t - rT) \tag{6.2}$$

and its frequency response is the Fourier transform of this, which is (from P1a and R6a)

$$K(f) = \sum_{r=-n}^{n} v_r \exp(-2\pi i r f T) \tag{6.3}$$

Thus, we can put

$$\begin{aligned} |K(f)G(f) - 1|^2 &= \left[\sum_{r=-n}^{n} v_r{}^* \exp(2\pi i r f T) G^*(f) - 1\right] \\ &\quad \left[\sum_{s=-n}^{n} v_s \exp(-2\pi i s f T) G(f) - 1\right] \qquad (6.4) \\ &= \sum_{r=-n}^{n} \sum_{s=-n}^{s} v_r{}^* v_s e^{2\pi i (r-s) f T} |G(f)|^2 \\ &\quad - 2 \operatorname{Re} \sum_{s=-n}^{s} v_r{}^* e^{2\pi i r f T} G(f)^* + 1 \end{aligned}$$

The error power that is to be minimized, as a function of the weight vector $\mathbf{v}$ (where $\mathbf{v} = [v_{-n} v_{-n+1} \ldots . v_n]^{\mathrm{T}}$), is given from (6.1), on substituting for $KG - 1$, from (6.4), by

$$p(\mathbf{v}) = \int_{-\infty}^{\infty} |K(f)G(f) - 1|^2 |U(f)|^2 \, df$$

$$= \sum_r \sum_s v_r{}^* v_s b_{rs} - 2 \operatorname{Re} \sum_r v_r{}^* a_r + c$$

or

$$p(\mathbf{v}) = \mathbf{v}^{\mathrm{H}}\mathbf{B}\mathbf{v} - 2 \operatorname{Re} (\mathbf{v}^{\mathrm{H}}\mathbf{a}) + c \tag{6.5}$$

where the components of **a** and **B** are given by

$$a_r = \int_{-\infty}^{\infty} G(f)^* |U(f)|^2 e^{2\pi i f r T} \, df \tag{6.6}$$

and

$$b_{rs} = \int_{-\infty}^{\infty} |G(f)|^2 |U(f)|^2 e^{2\pi i f (r-s) T} \, df \tag{6.7}$$

and c is $\int |U(f)|^2 \, df$. (We can normalize the error power relative to the signal power by dividing by c or, equivalently, by normalizing U so that $c = 1$; we will take this to be the case.) We note that (6.5) and (6.6) are in the form of (inverse) Fourier transforms. If $\rho_1(t)$ and $G(f)^*|U(f)|^2$ are a Fourier pair and so are $\rho_2(t)$ and $|G(f)|^2|U(f)|^2$, then from (6.6) and (6.7) we have

$$a_r = \rho_1(rT) \text{ and } b_{rs} = \rho_2[(r-s)T] \tag{6.8}$$

As in Section 5.3, we differentiate p in (6.5) with respect to $\mathbf{v}$ to find that the mismatch error is minimized at $\mathbf{v}_0$, given by

$$\mathbf{v}_0 = \mathbf{B}^{-1}\mathbf{a} \tag{6.9}$$

and the minimum (normalized) squared error is

$$p(\mathbf{v}_0) = 1 - \mathbf{a}^{\mathrm{H}}\mathbf{B}^{-1}\mathbf{a} = 1 - \mathbf{a}^{\mathrm{H}}\mathbf{v}_0 \tag{6.10}$$

(We note that $\mathbf{a}^{H}\mathbf{B}^{-1}\mathbf{a}$ is real as, from (6.7), **B** is Hermitian, i.e., $b_{sr} = b_{rs}^{*}$.) Thus, in order to find the optimum tap weights (in the sense of giving least squared error) for the equalizing filter, we need only $|U|^2$, the power spectrum of the signal, and G, the complex channel response, and then we perform the Fourier transforms defined in (6.6) and (6.7) to give the components of **a** and **B**, followed by some simple matrix processing. The derivation of **a** and **B** in the case of a simple delay mismatch has been given in Section 5.3, but the cases of frequency-dependent amplitude mismatches as well are considered in Sections 6.4 and 6.7 below. The delay mismatch is a linear phase dependence on frequency, but we do not go on to cover the case of nonlinear phase correction, as the Fourier methods illustrated here are not so convenient for handling phase functions rather than (normally real) amplitude functions.

6.3 ramp and snc$_r$ Functions

Although the function G, describing the channel frequency response to be compensated, can be defined over the whole frequency domain, we are only interested in its form in the frequency interval containing significant signal energy. If, as we have generally assumed, the signal is limited (after down-conversion to complex baseband) to the band $(-F/2, F/2)$, then it will make no difference in the Fourier transform integrals of (6.6) and (6.7) if the function $\operatorname{rect}(f/F)$ is included, as the factor $|U(f)|^2$ is taken to be zero anyway in the region where this rect function is zero. Thus, if we consider first the case where G is a linear function of frequency, to avoid the problem of the function $G(f) = af + b$ being unbounded as $f \to \pm\infty$, we can take, more conveniently, $G(f) = (af + b)\operatorname{rect}(f/F)$. In order to handle polynomial functions of this kind, we introduce the function ramp defined by

$$\operatorname{ramp}(x) = 2x \operatorname{rect}(x) \tag{6.11}$$

and this is illustrated in Figure 6.2.

Thus $\operatorname{ramp}(x) = 2x$ on $-1/2 < x < 1/2$, and $\operatorname{ramp}(x) = 0$ for $x < -1/2$ and $x > 1/2$. [If required, we can take $\operatorname{ramp}(\pm 1/2) = \pm 1/2$.] As the rect function has the property $\operatorname{rect}^r(x) = \operatorname{rect}(x)$, we see that

$$\operatorname{ramp}^r(x) = (2x)^r \operatorname{rect}(x) \tag{6.12}$$

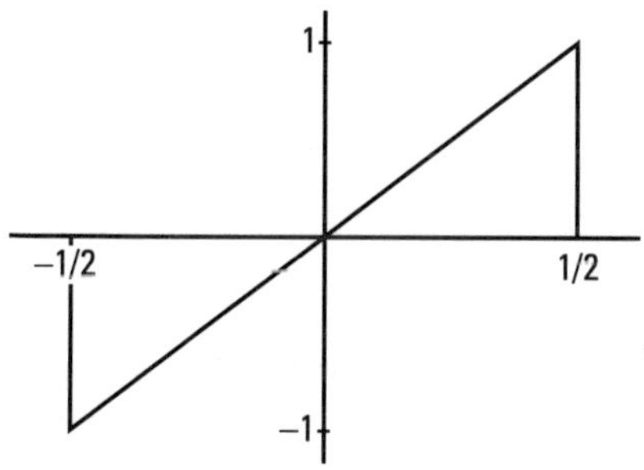

Figure 6.2 The ramp function.

so that we can express a polynomial in x on the interval (−1/2, 1/2) as a polynomial in ramp (x):

$$(a_0 + a_1 x + a_2 x^2 + \ldots)\,\mathrm{rect}\,(x) = a_0\,\mathrm{ramp}^0\,(x) + (a_1/2)\,\mathrm{ramp}\,(x) + (a_2/4)\,\mathrm{ramp}^2\,(x) + \ldots \quad (6.13)$$

To find the Fourier transform of ramp, we use Rule R9b:

$$-2\pi i x u(x) \Leftrightarrow U'(y) \quad (6.14)$$

where $u(x) \Leftrightarrow U(y)$ and the prime denotes the derivative. If we define $V(y)$ as $U'(y)$, with (inverse) Fourier transform $v(x)$, then, from (6.14), $v(x) = -2\pi i x u(x)$ and also, by Rule 9b, $-2\pi i x v(x) \Leftrightarrow V'(y)$. Substituting for v and V gives

$$(-2\pi i x)^2 u(x) \Leftrightarrow U''(y)$$

and, in general, for any positive integer r,

$$(-2\pi i x)^r u(x) \Leftrightarrow U^{(r)}(y) \quad (6.15)$$

where $U^{(r)}$ is the rth derivative of U. Now putting $u(x) = \mathrm{rect}\,(x)$ and $U(y) = \mathrm{sinc}\,(y)$, from Pair 3a, then substituting in (6.15), we obtain

$$(-\pi i)^r (2x)^r\,\mathrm{rect}\,(x) \Leftrightarrow \mathrm{sinc}^{(r)}\,(y) \quad (6.16)$$

If we introduce the notation

$$\mathrm{snc}_r(y) = \frac{1}{\pi^r}\frac{d^r}{dy^r}\,\mathrm{sinc}(y) \tag{6.17}$$

then, from (6.12), (6.16) becomes

$$\mathrm{ramp}^r(x) \Leftrightarrow i^r\,\mathrm{snc}_r(y) \tag{6.18}$$

We note from (6.11) and (6.16) that we can write, formally,

$$\mathrm{ramp}^0(x) = \mathrm{rect}(x) \text{ and } \mathrm{snc}_0(y) = \mathrm{sinc}(y) \tag{6.19}$$

From (6.17), carrying out the differentiation, we find

$$\mathrm{snc}_1(y) = \frac{\cos(\pi y) - \mathrm{snc}_0(y)}{\pi y} \qquad (n \geq 1) \tag{6.20}$$

This holds for all real values of y except for $y = 0$, so we define $\mathrm{snc}_1(0) = 0$ to ensure that snc_1 is continuous and, in fact, analytic. Differentiating again, we obtain

$$\mathrm{snc}_2(y) = -\mathrm{snc}_0(y) - \frac{2\,\mathrm{snc}_1(y)}{\pi y} \tag{6.21}$$

with $\mathrm{snc}_2(0) = -1/3$. These three functions have been plotted in Figure 6.3. We note that the even order snc functions are even functions and the odd ones are odd functions. The maximum level of the functions falls as the order rises.

We note that (6.21) [unlike (6.20)] contains all the trigonometric functions in snc functions only. By differentiating further, using (6.17) we can obtain a recursion formula, of which (6.21) is the first example, from which higher order snc functions can be found:

$$\mathrm{snc}_n(y) + \mathrm{snc}_{n-2}(y) = -\frac{n\,\mathrm{snc}_{n-1}(y) + (n-2)\,\mathrm{snc}_{n-3}(y)}{\pi y} \qquad (n \geq 2) \tag{6.22}$$

By expressing $\sin(\pi x)$ in its Taylor series form and differentiating term by term for the two higher order functions, we find, for the first three snc functions,

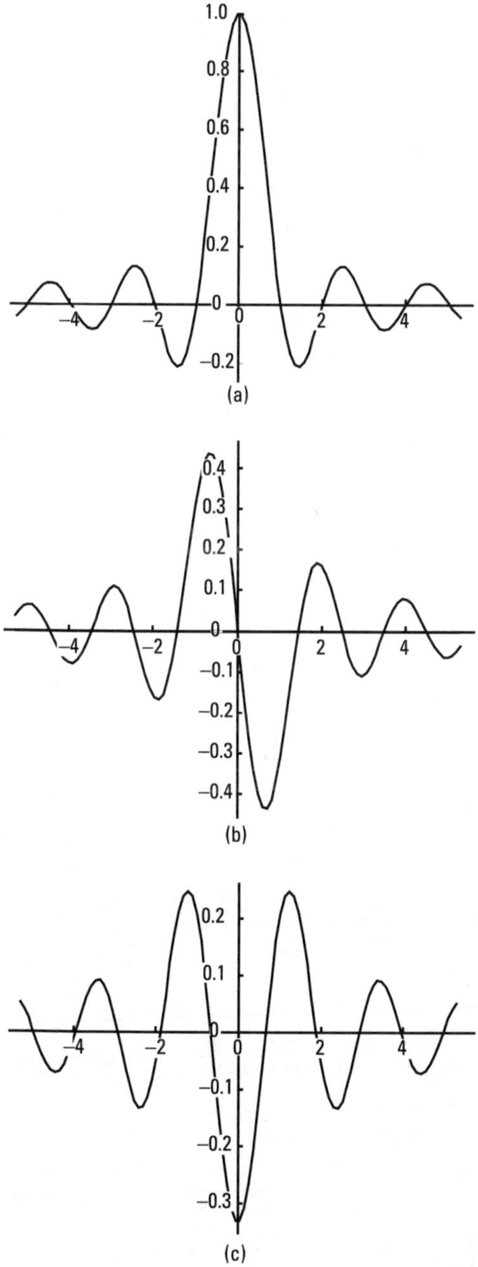

Figure 6.3 First three snc functions: (a) snc_0; (b)snc_1; and (c) snc_2.

$$\mathrm{snc}_0(y) = \sum_{n=0}^{\infty} \frac{(-1)^n (\pi y)^{2n}}{(2n+1)!} \tag{6.23}$$

$$\mathrm{snc}_1(y) = \sum_{n=0}^{\infty} \frac{(-1)^n 2n(\pi y)^{2n-1}}{(2n+1)!} = \sum_{n=1}^{\infty} \frac{(-1)^n 2n(\pi y)^{2n-1}}{(2n+1)!} \tag{6.24}$$

$$\mathrm{snc}_2(y) = \sum_{n=0}^{\infty} \frac{(-1)^n 2n(2n-1)(\pi y)^{2n-2}}{(2n+1)!} = \sum_{n=1}^{\infty} \frac{(-1)^n 2n(2n-1)(\pi y)^{2n-2}}{(2n+1)!} \tag{6.25}$$

In general, we can put

$$\mathrm{snc}_r(y) = \sum_{n=[(r+1)/2]}^{\infty} \frac{(-1)^n 2n!(\pi y)^{2n-r}}{(2n-r)!(2n+1)!} \tag{6.26}$$

where $[p]$ is the highest integer in p, so $[(r+1)/2] = (r+1)/2$ for r odd and $[(r+1)/2] = r/2$ for r even. The even order series contains only even powers of y and so are even functions, and the odd series contains only odd powers and are odd functions. Thus, for all the odd order snc functions, we have $\mathrm{snc}_r(0) = 0$, while from (6.26) we see that for r even, say $r = 2s$, when $y = 0$ the only nonzero term is the first, for which $n = r/2 = s$, so that

$$\mathrm{snc}_{2s}(0) = \frac{(-1)^s 2s!}{0!(2s+1)!} = \frac{(-1)^s}{2s+1} \tag{6.27}$$

6.4 Simple Example of Amplitude Equalization

It has already been remarked that the problem of delay equalization as considered here has been covered in Section 5.3 under the subject of sampled waveform delay, so no further illustrations are given here. However, the subject of amplitude equalization has not been illustrated before, so a simple example, using the results of Section 6.3 above, is presented in this section, showing how effective the method is and with how little computation if there is a degree of oversampling. We take the simple case of a linear amplitude distortion with an unweighted squared error function over the bandwidth (equivalent to a rect function power spectrum). The response to be matched is of the form $G(f) = 1 + af$ over the bandwidth (taken to be unity), or $G(f) = \text{rect}\, f + (a/2)\, \text{ramp}\, f$. The Fourier transforms of G required for the components of **a** will include a transform of the ramp function, that is, a snc_1 function, as well as a snc_0 from the rect function. As we require the transform of $G^2(f)$ to determine the elements of **B**, we also have a ramp^2 function with its transform snc_2. There is an important detail to notice in that they are actually *inverse* Fourier transforms that are required [see (6.6) and (6.7)]. In many cases (using symmetric functions, in particular), there is no distinction between forward and inverse transforms, but here we have odd functions (ramp and snc_1). We see from (6.18) that ramp^r (forward) transforms to $i\ \text{snc}_r$, so, from Rule 4, we have $i\ \text{snc}_1(x) \Leftrightarrow \text{ramp}(-y) = -\text{ramp}(y)$ (as ramp is odd)—that is, $\text{ramp}(f)$ inverse transforms to $-i\ \text{snc}_1(t)$. (However, as ramp^2 is even, it transforms to $+i^2$ snc_2 in both the forward and inverse cases.)

For Figure 6.4 we have taken a linear amplitude distortion for $G(f)$ of 10 dB across the band, from an amplitude of 0.48 to 1.52. Using a seven-element equalizing filter and a relative sampling rate of 1 (no oversampling), we get a useful degree of equalization [Figure 6.4(a)]. The filter response K, which should ideally be the reciprocal of G over the band, is shown, as well as the equalized response KG. If we increase the oversampling rate to 1.5, or 50% oversampling, the equalization becomes very good [Figure 6.4(b)]. To get a comparable ripple performance at the basic sampling rate, we see that we have to increase the number of filter taps greatly—even at 47 taps [Figure 6.4(c)] the ripples are greater; the higher ripple frequency is due to the much greater time spread of the taps in this case.

Finally, for Figure 6.4(d) there was both amplitude variation and delay to be compensated. The same linear amplitude function was taken, with a delay error of 0.5 sampling interval, and the filter parameters are the same as in Figure 6.4(b). In this case the functions have some residual phase variation, so the modulus has been plotted, and we see that this has been very well equalized within the band—almost identically with the case of no delay error—but varies significantly (particularly on the positive frequency side) outside the band.

6.5 Equalization for Broadband Array Radar

Many antennas for use in radio, radar, or sonar systems consist of an array of simple elements, rather than, in some radio cases, a single element or,

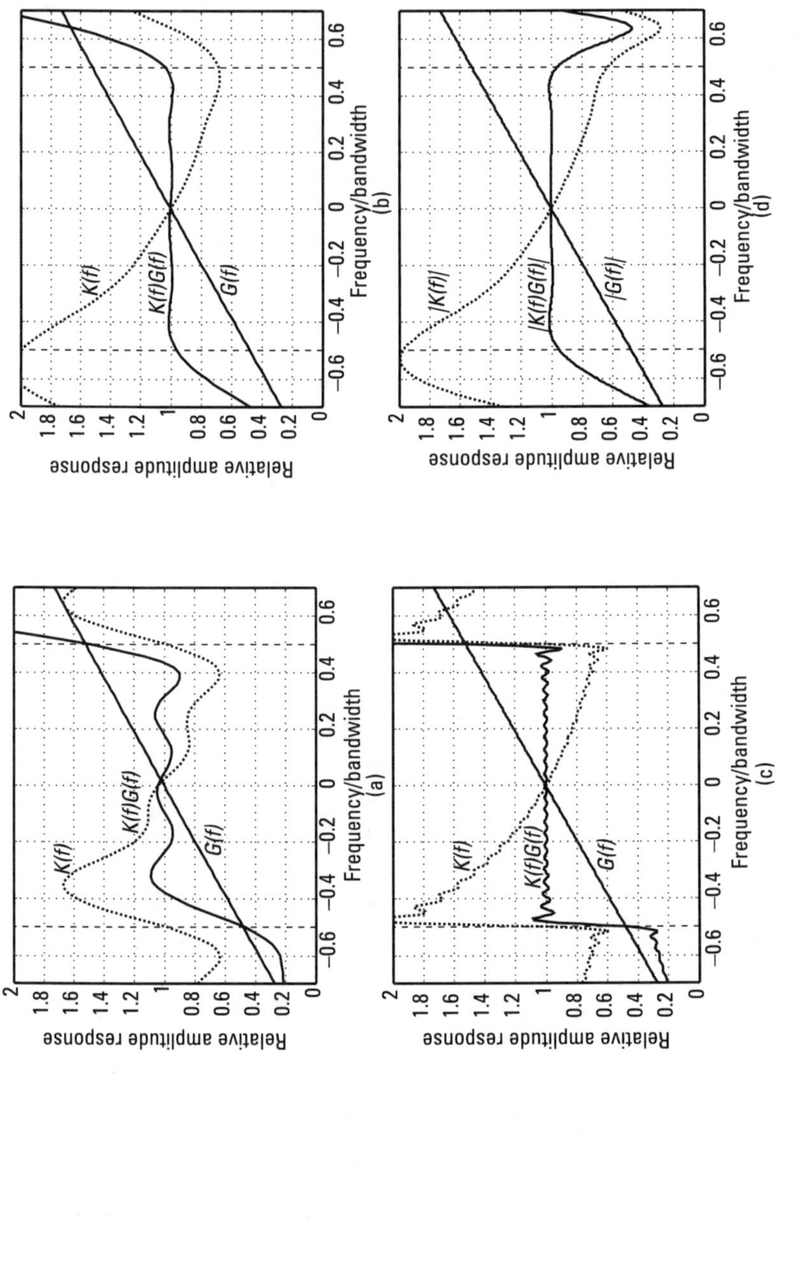

Figure 6.4 Equalization of linear amplitude distortion: (a) $m = 7$, $q = 1$; (b) $m = 7$, $q = 1.5$; (c) $m = 47$, $q = 1$; (d) $m = 7$, $q = 1.5$, delay 0.5.

for radar and satellite communications, a large parabolic dish or even an exponential horn. For maximum signal-to-noise ratio (whether on transmission or reception) in a particular direction, the signals passing through the elements must be adjusted in phase so that they sum in phase at the frequency of operation. Of course, in practice all signals occupy a finite bandwidth, so that in principle different phase shifts are needed across this band, as it is really a time difference, dependent on element position, that needs compensation. However, many signals are narrowband in that the fractional bandwidth, the ratio of the bandwidth to the carrier frequency, is small. In this case the phase shift required across the band is close to that at the center frequency, and as it is much easier to apply a simple (frequency-independent) phase shift than a delay, this approximation can be used. Whether this approximation is acceptable or not in a given system depends not only on the fractional bandwidth, but also on the size, or aperture, of the array. Thus, *narrowband* is a relative term, and perhaps the most appropriate definition of a narrowband signal in this context is that it can be termed narrowband if ignoring its finite bandwidth leads to negligible, or practically acceptable, errors. Conversely, a broadband signal as defined here is one where this is not the case, and allowance, or compensation, must be made for the different frequencies across its bandwidth to maintain the required performance. (There seems to be no standard definition of these terms, but this qualitative definition seems to be clearer in some ways than a quantitative one; for a very small array, a 5% band may be "narrow" in this sense, while a 1% band may be "broad" in the context of a very large and hence highly frequency-sensitive, aperture. We will use *wideband* for the case where the band of interest extends down to 0 Hz; this is the same as the 200% broadband case and is consistent with the use of the term in Section 4.3.)

The problem is illustrated in Figure 6.5 for a simple linear array. An element at distance d from the center of the array receives the signal from direction θ, relative to broadside, at time τ earlier than at the reference point, where

$$\tau(\theta) = d \sin(\theta)/c \tag{6.28}$$

where c is the velocity of light. Thus, in principle, the output of this element should be delayed by $\tau(\theta)$ to steer the array in direction θ, but as phase shifts are much more easily implemented than delays, it is usual, using the narrowband condition, to introduce the phase shift

$$\phi(\theta) = 2\pi f_0 \tau(\theta) = 2\pi(d/\lambda_0)\sin\theta \tag{6.29}$$

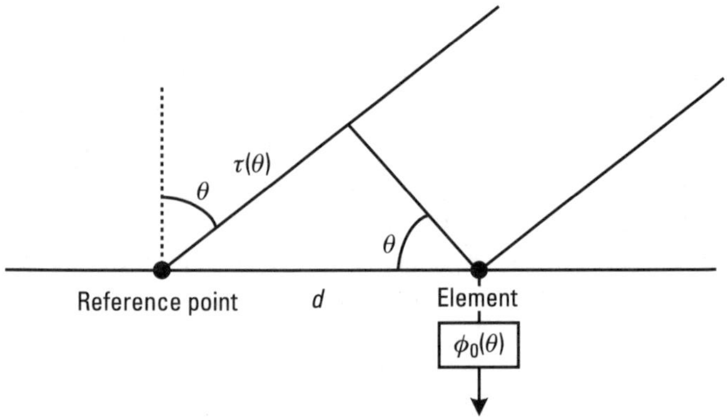

Figure 6.5 Array steering.

where f_0 is the center frequency, λ_0 the corresponding wavelength (such that $f_0\lambda_0 = c$), and d/λ_0 is the distance of the element from the reference point in wavelengths. [More generally, if the element position vector is $\mathbf{r}$ and the unit vector in the direction of interest is $\mathbf{e}(\alpha, \epsilon)$ for azimuth α and elevation ϵ, then the required phase shift to steer in direction (α, ϵ) is $2\pi\mathbf{r}.\mathbf{e}(\alpha, \epsilon)/\lambda_0$, where $\mathbf{r}.\mathbf{e}$ is the scalar product of these vectors].

Summing the element outputs in phase produces the peak response in the steered direction, and this form of response is known as the *sum beam.* (Strictly, this is only the array factor; for the full response this is multiplied by the element response, in the case of essentially identical elements.) For high angular accuracy in radar, a technique known as *monopulse measurement* is used, which requires a *difference beam,* which ideally has zero gain in the look direction and a linear amplitude response near this direction. The angular offset of a target from the look direction is found by observing the level of its echo in the difference beam (normalized by the sum beam response) and dividing by the known slope of this beam. One form of difference beam, in the case of a regular linear or planar array, is obtained by dividing the array into two equal parts and subtracting the responses of the two halves (hence the origin of the name), but an alternative approach, which allows a difference beam to be formed with a more general geometry, is to form a beam that is the angular derivative of the sum beam. This is the form that will be considered in Section 6.7.

6.6 Sum Beam Equalization

To steer a broadband sum beam, we need only to replace the simple phase shift at the carrier frequency corresponding to the relative delay that is to be compensated, which is applied to the output of each element in the narrowband case, with the delay itself. In fact, it is not easy in practice to provide flexible delays at RF, as would be required for a beam to be steered freely in various directions, but for arrays with digital processing, we can provide a very close approximation to the required delays using the methods discussed here, which can be implemented rapidly. In fact, as the processing is carried out at baseband, after downconversion and digitization, the delay is implemented at baseband. The phase shift on the carrier is still required and can be applied at the RF stage, as for the narrowband application, or digitally after downconversion, but is independent of the equalization process. For the sum beam, considered here, the channel equalization is a simple delay, which can be approximated by the methods of Chapter 5, so this application of equalization requires essentially no new ideas. However, we show the benefit of this equalization in the example of a simple array, and in the next section we consider the equalization for the difference beam, which is more complex and uses the results of Sections 6.2 and 6.3, and we illustrate this using the same array.

The system we model is a 16-element linear array of omnidirectional elements at half-wavelength spacing operating at 3,000 MHz. (More generally, the elements need not be omnidirectional, but should be similar and are taken to be frequency-independent.) The actual frequency is not particularly significant; more important are the relative frequencies. To illustrate the problem, Figure 6.6 shows the gain (more precisely, the array factor) at three frequencies for the array when steered at 50 degrees from broadside. The steering weights are the correct phases at the center frequency, and this beam has its peak at the correct position. With the same weights, the beams at the frequencies 200 above and 200 MHz below (at about ±6.7% offset) are displaced in position. (This effect is known as *squint*.) Thus, for a broadband signal arriving from 50 degrees and with the array steered in this direction, there will be a variation in gain, about 2½ dB over the 400-MHz band in this case, and hence a distortion of the received signal.

Using the equalization method described in Section 6.2, we use (6.9) for the weights on the FIR filter taps, where the components of **a** and **B** are given in general form in (6.6) and (6.7). For this example we take a trapezoidal shape for the spectral power density *U*, and the channel response requiring compensation, or equalization, *G* is just that due to a simple delay (different in each element channel, in general). For this delay equalization case, the problem is the same as that considered in Section 5.3, and the components of **a** and **B** are given more specifically by (5.46a, b). The result of implementing this delay equalization is shown in Figure 6.7. The receiver

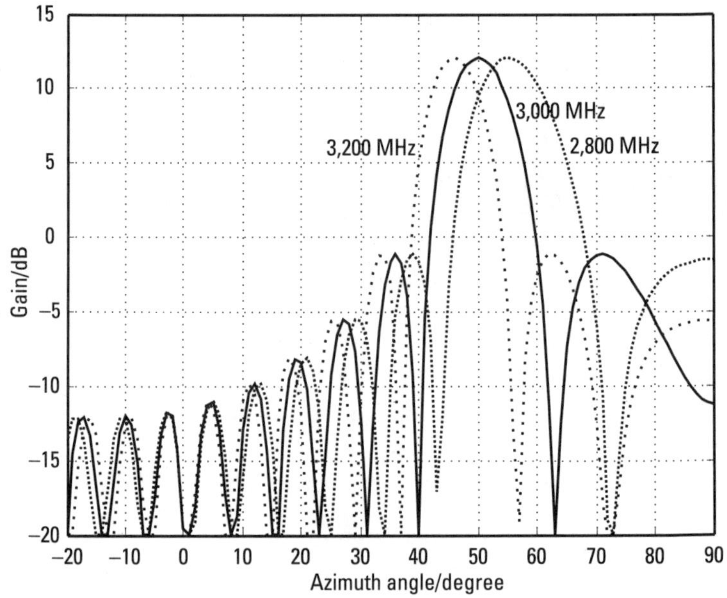

Figure 6.6 Array response with narrowband weights.

bandwidth was taken to be 500 MHz, so the frequencies of the responses at ±200 MHz are towards the edges of the band, and we see that the squint has been effectively removed, with the main lobes of the response at the three frequencies virtually coincident. Two processing alternatives are shown; in Figure 6.7(a) only five taps are used for each delay with an oversampling rate of 1.2, or 20% above the minimum rate, and in Figure 6.7(b) only four taps are used, but the sampling rate increase has been raised to 50%.

To study the response at the peak of the beam more closely, the gain in the steered direction has been plotted in Figure 6.8 as a function of frequency normalized to the center frequency. The plot in Figure 6.8(a) is for the case of five taps and a 20% oversampling rate, as in Figure 6.7(a), and it can be seen that there is only a slight variation with frequency—a rise of less than 0.2 dB and a fall in gain at the very edges of the band where the signal power density is falling and the matching is not required to be so good. In this figure, the vertical lines mark the edge of the 10% band over which equalization is required, and the dashed curve shows the response in the absence of equalization. In Figure 6.8(b), the parameters of the equalization filters are the same (five taps and 20% oversampling), but the receiver bandwidth is now 200%, extending from zero to twice the center frequency. Again, the dashed line shows the frequency response without any

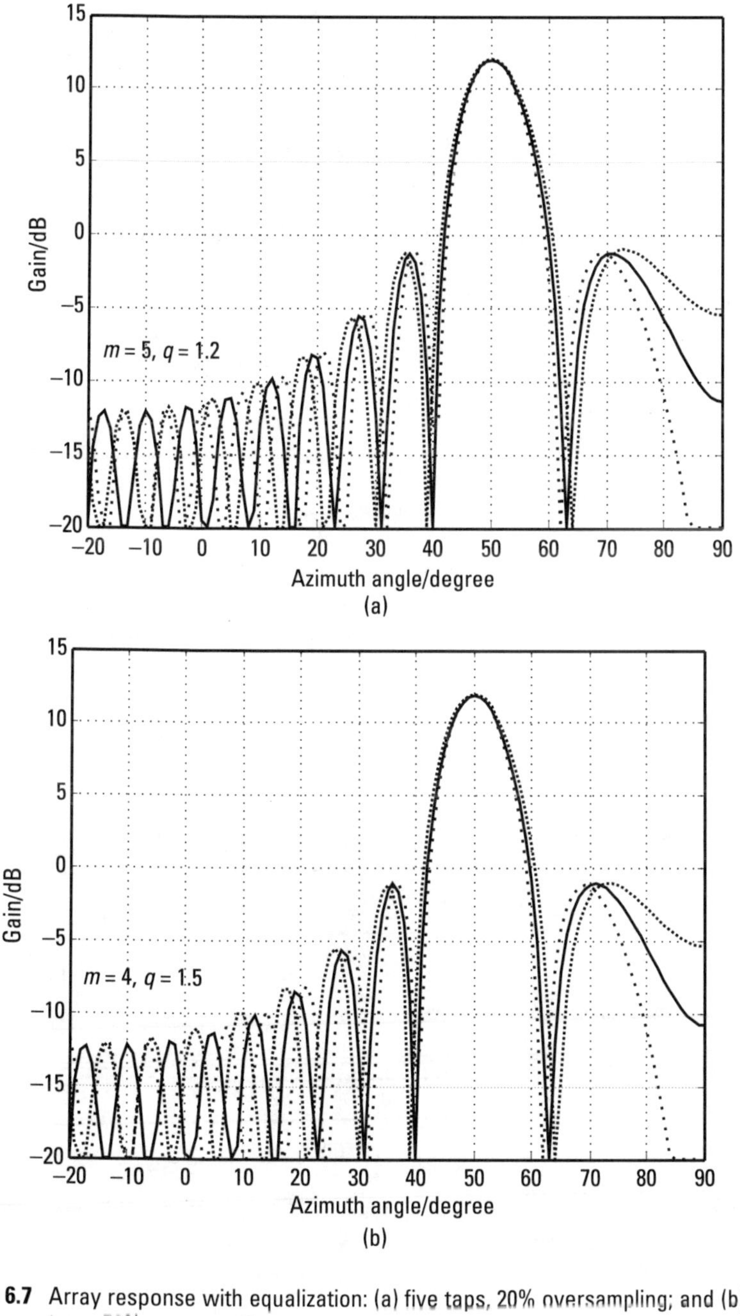

Figure 6.7 Array response with equalization: (a) five taps, 20% oversampling; and (b) four taps, 50% oversampling.

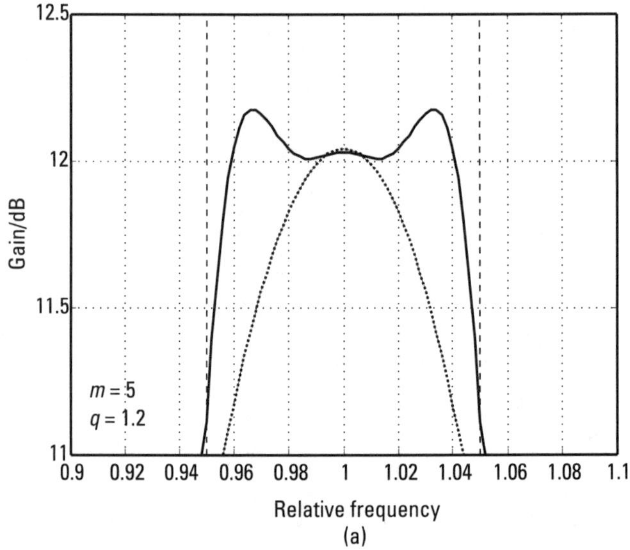

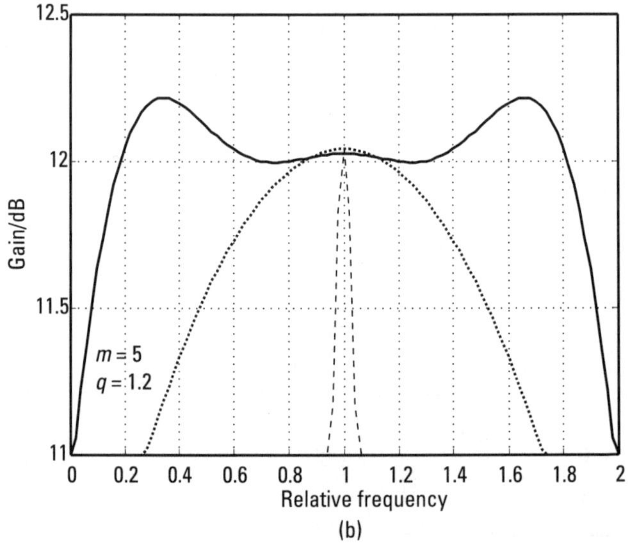

Figure 6.8 Sum beam frequency response; effect of bandwidth: (a) 10% bandwidth; and (b) 200% bandwidth.

equalization, and the dotted response is that for the case of only integer delay compensation (in units of the sampling interval). [In Figure 6.8(a), this curve is identical to the dashed curve, as all the delays are within ±0.5 sampling intervals, so no integer compensation is feasible.] We note that, with the same set of parameters, the response is essentially independent of the fractional bandwidth—the shapes of the responses are virtually identical. In the second case, the sampling rate is much higher, of course; in this case, as the bandwidth is $2f_0$, where f_0 is the center frequency, the sampling rate is $2.4f_0$. To sample at this rate may be impracticable at radar frequencies, but may well be feasible for sonar, where broadband (or even wideband) operation is much more commonly required and the actual signal frequencies are much lower.

The fact that these responses are very similar is not a coincidence, but illustrates the fact that the response is essentially independent of the fractional bandwidth and only depends on how well the delays are matched. This depends, for a given set of equalization filter parameters (m and q), on how close the required delays are to integer multiples of the sampling period. This will vary, in general, from one element to another and will depend on the beam steered direction and the element separation. In particular cases, the delays required may all be integral, in sampling periods, in which case the matching will be exact, in principle, and the response will be completely flat. At the other extreme, the delays required might all be half-integral, which is the worst case for matching. In general, however, there will be a spread of delays and the performance will be intermediate. This is illustrated in Figure 6.9. Here the frequency axis is the frequency offset from the center, normalized to the bandwidth, so the range shown is just the band over which equalization is required. For this figure the parameters were chosen in order to include the two extreme cases described above. The delay required for an element at distance d from the array center is given in (6.28), and putting $c = f_0 \lambda_0$ and dividing by the sampling period $1/qF$, the required delay in units of sampling periods is given by

$$\beta = (d/\lambda_0)\, q \sin\theta\, (F/f_0) \tag{6.30}$$

The element separation was increased to one wavelength, so that the element positions are given as $(2n + 1)/2$ wavelengths (n an integer from −8 to +7 for the 16-element array). The steer direction remained at 50 degrees, but q was increased to 1.3054 so that $q \sin 50° = 1$. The delay required for element n is then given, from (6.28), by $(2n + 1)(F/2f_0)$. If we choose $F = f_0$, the 100% bandwidth case, we see that all the delays are

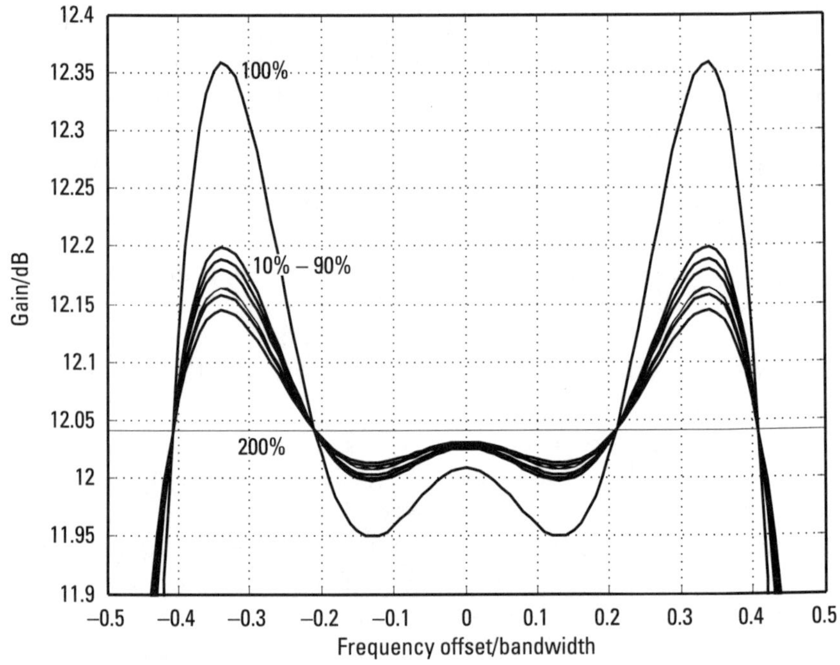

Figure 6.9 Sum beam response with frequency offset for various bandwidths.

half integral, the worst case, while if $F = 2f_0$, the 200% case, the delays are integral and we have the flat response shown. The other curves are for the cases of 10%, 20%, . . . , 90% bandwidths (*not* giving a monotonic sequence of peak ripple levels), and because the fractional delays are distributed over the full range, the results are much the same and intermediate between the extreme cases. These are also the same results as for bandwidths of 110% to 190%, because it can be seen that for this case (where $q \sin \theta = 1$), the results for a fractional bandwidth F/f_0 of r and $2 - r$ will give the same result.

The effect of varying the equalization parameters is shown in Figure 6.10 for the 50% bandwidth case and for the array separation of 0.5 wavelengths. In Figure 6.10(a), we fix the sampling rate at 1.2, or 20% above the minimum, and vary the number of taps m in the equalizing filters. We note that even with three tap filters, the ripple in the center of the band is quite small (up to 0.25 dB above the fully equalized level), but there is a rather rapid fall in gain at the edges of the band, starting well within the flat-top region (at ±0.4 bandwidths offset). As m increases, the response improves, and at $m = 9$ the gain only falls off rather sharply outside the flat top of the signal

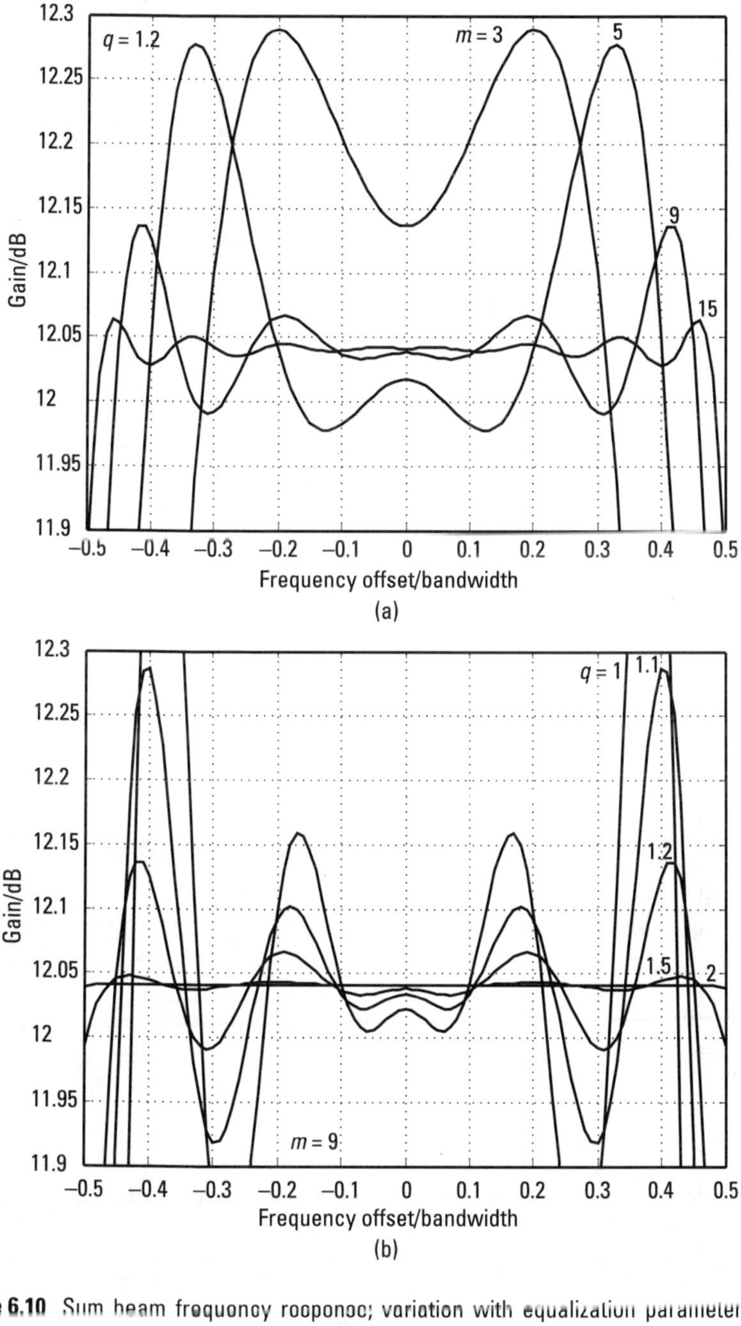

Figure 6.10 Sum beam frequency response; variation with equalization parameters: (a) variation with m; and (b) variation with q.

spectrum. The filter length has been kept at 9 in Figure 6.10(b), and the relative sampling rate q is varied. With no oversampling ($q = 1$), the nine tap filters do achieve a considerable degree of equalization, but there is a large ripple near the edges of the band. This is rapidly reduced with oversampling, and at $q = 2$ the response is almost perfect. (It should be remarked that these are nominal gain plots and give an ideal figure of 12.04 dB for an array of 16 elements. They should be corrected slightly if directivity is required, but any correction will generally be small, particularly for larger arrays not steered too near to a grating lobe condition.)

Finally, Figure 6.11 shows clearly the benefit of oversampling. At the minimum sampling rate, a very long filter is needed for effective equalization—in this example, 101 elements are required (continuous curve) to give low ripples in the response. If the sampling rate is increased to just 1.1, comparable ripples result at a filter length of only 21 (dashed curve), a reduction of nearly five times in the computation required. Oversampling at 50% (dotted curve) allows an improvement by a further factor of three to only seven elements. For planar arrays with a large number of elements,

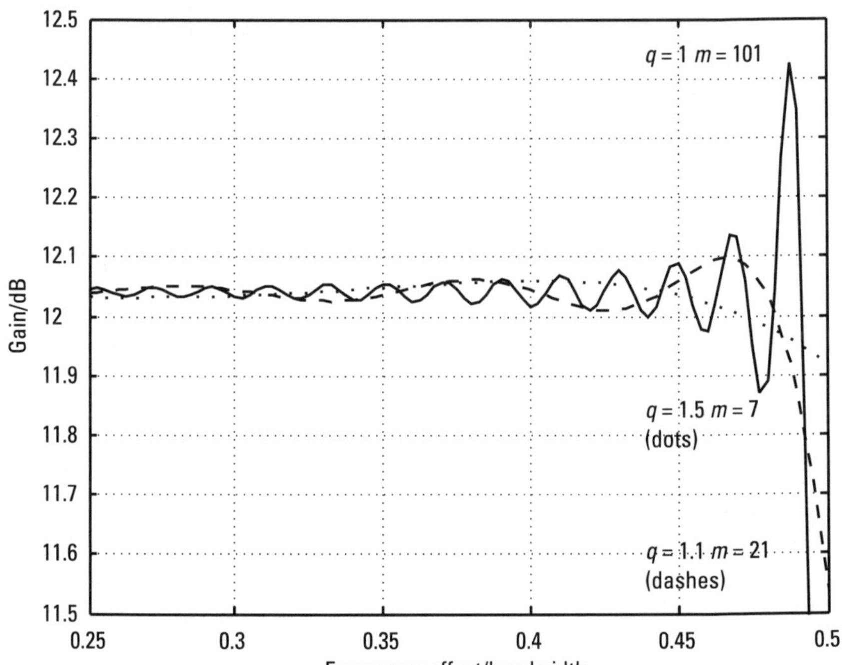

Figure 6.11 Effect of increasing oversampling rate.

typically required for many radars, it could be important and valuable to keep the complexity of equalization down to a modest level in each channel. In some applications, with a moderate degree of oversampling, filters of length as low as three or four may be adequate.

6.7 Difference Beam Equalization

We take the difference beam pattern to be given by the derivative with respect to the angle of the sum beam pattern. We will use the *sine-angle* coordinate u, where $u = \sin\,\theta$, as this simplifies the expressions below, particularly for the difference beam slope, but otherwise does not affect the principles being illustrated. (In this form, the beam shape, plotted against u, remains unchanged as the beam is scanned.) Thus, in this section, we replace $\sin\theta$ with u, particularly in equations that use (6.28). If $w_k(u_0)$ is the weight applied to the output of element k to steer in direction θ_0, where $u_0 = \sin\,\theta_0$, then the sum beam gain (array factor) is given, as a function of frequency and angle, by

$$g(u, f;\, u_0) = \sum_k w_k\,(u_0) \exp\, 2\pi i f \tau_k(u) \tag{6.31}$$

For narrowband steering we take $w_k(u_0) = \exp(-2\pi i d_k f_0 u_0/c)$, so that the signals add in phase in the look direction θ_0 at the center frequency f_0. The sum is over all elements and the signal delay τ_k relative to the center of the array is given by $d_k u/c$ [from (6.28)], where the element is at distance d_k from the array centroid, the mean element position, such that the sum of the element positions measured from this point is zero. The difference beam pattern is therefore given, with the prime indicating differentiation with respect to sine-angle, by

$$h(u, f;\, u_0) = g'(u, f;\, u_0) = \sum_k w_k(u_0)(2\pi i f d_k/c) \exp\, 2\pi i f d_k u/c \tag{6.32}$$

In fact, to form a difference beam we do not need the factor f; we only require the frequency-sensitive element delay compensation factors $w_k(\theta_0) = \exp(-2\pi i f d_k u_0/c)$, which allow the signals to sum in phase across the frequency band. The element distances d_k are weighting factors that result in zero gain in the look direction with these weights applied. The weights w_k are the same as required for the sum beam, so the same frequency compensation is required on each element. Thus, excluding the factor f in (6.32) and also factors independent of frequency, we define the difference beam response, within a scaling factor, by

$$h(u, f;\, u_0) = \sum_k w_k(\theta_0)\, d_k \exp 2\pi i f d_k u/c \tag{6.33}$$

However, for an ideal difference beam, we require its *slope,* with respect to angle, at the beam pointing position θ_0 to be constant across the band, and this is the derivative of h with respect to angle:

$$h(u, f;\, u_0)' = -\sum_k w_k(\theta_0)\, d_k\, 2\pi f (d_k/c) \exp 2\pi i f d_k u/c$$

In this case we cannot remove the variable f from the expression, because this is not a definition of the slope, but a derivation from the pattern as defined in (6.33). Omitting the constant $2\pi/c$, we have for the difference beam slope, within a scaling factor,

$$s(u, f;\, u_0) = -\sum_k w_k(\theta_0)\, d_k^2 f \exp 2\pi i f d_k u/c \tag{6.34}$$

In (6.34) f is a frequency within the RF band (i.e., $f_0 - F/2 < f < f_0 + F/2$), but if we now want to represent the gain pattern in terms of the baseband frequency, we replace f with $f_0 + f$, where now we have $-F/2 < f < F/2$. With this change, the response at baseband frequency f, after down-conversion (which removes f_0 from the exponential factor) is given by

$$s(f, u;\, f_0, u_0) = -\sum_k w_k(u_0)(f_0 + f)\, d_k^2 \exp 2\pi i f d_k u/c$$

Dividing by f_0 and then removing this scaling factor, we have

$$s(f, u;\, f_0, u_0) = -\sum_k w_k(u_0)(1 + f/f_0)\, d_k^2 \exp 2\pi i f d_k u/c \tag{6.35}$$

This response varies with angle and frequency, but we require it to be independent of frequency at the direction of interest θ_0. Thus, excluding constants with respect to frequency, we see that the frequency variation to be compensated is now of the form

$$S(f) = (1 + \phi f/F) \exp 2\pi i f \tau_k(u_0) \tag{6.36}$$

where the delay τ_k varies with the element position, and we have expressed the function S in terms of ϕ, the fractional bandwidth F/f_0.

Before putting this expression for S into (6.6) and (6.7), we note that for the band-limited signal we effectively have a factor $\text{rect}(f/F)$ in $|U(f)|^2$, so multiplying $S(f)$ by this rect function will make no difference to the integrals in (6.6) and (6.7). Thus, we can replace S by

$$S(f)\,\text{rect}\left(\frac{f}{F}\right) = \left[\text{rect}\left(\frac{f}{F}\right) + \frac{\phi}{2}\,\text{ramp}\left(\frac{f}{F}\right)\right] \exp(2\pi i f \tau) \tag{6.37}$$

Putting this into (6.6) and (6.7) (in place of G) gives

$$a_r = \int_{-\infty}^{\infty} \left[1 + \frac{\phi}{2}\,\text{ramp}\left(\frac{f}{F}\right)\right] |U(f)|^2 \exp 2\pi i f(rT - \tau)\, df \tag{6.38}$$

and

$$b_r = \int_{-\infty}^{\infty} \left[1 + \phi\,\text{ramp}\left(\frac{f}{F}\right) + \frac{\phi^2}{4}\,\text{ramp}^2\left(\frac{f}{F}\right)\right] \tag{6.39}$$

$$|U(f)|^2 \exp[2\pi i f(r - s)T]\, df$$

Now let ρ_a, ρ_b, and ρ_c be the Fourier transforms of $|U(f)|^2$, $\text{ramp}(f/F)|U(f)|^2$, and $\text{ramp}^2(f/F)|U(f)|^2$, respectively, and also let us put $\tau = (p + \beta)T$, where $-0.5 < \beta \le 0.5$ and p is integral. As before, we assume that the delays are compensated to the nearest integer multiple p of the sampling period by taking the appropriate sampled pulse train (from a shift register, for example) and that we only have to equalize the fractional parts using the filter. Introducing these, we see that (6.38) and (6.39) can be written

$$a_r = \rho_u[(r - \beta)T] + (\phi/2)\,\rho_b[(r - \beta)T] \tag{6.40}$$

and

$$b_r = \rho_a[(r - s)T] + \phi\rho_b[(r - \beta)T] + (\phi^2/4)\rho_c[(r - s)T] \tag{6.41}$$

Now, for the trapezoidal spectrum we have [as in (5.44)]

$$|U(f)|^2 = \frac{4}{(1 + a)(1 - a)F^2} \operatorname{rect}\left[\frac{2f}{(1 - a)F}\right] \otimes \operatorname{rect}\left[\frac{2f}{(1 + a)F}\right] \tag{6.42}$$

[Although the function $\operatorname{rect}(f/F)$ does not appear in this expression, the spectral function would be unchanged on multiplying by this rect function, as the convolution of the rect functions in (6.42) has a base width of $(1 - a)F/2 + (1 + a)F/2 = F$, the same as $\operatorname{rect}(f/F)$. The rect function is unity within the region where the trapezoidal function is nonzero and zero where the trapezoidal function is zero. This justifies the statement above (6.37) that this rect function can be included in the integral and hence also with S.]

The Fourier transform of the power spectrum in (6.42) is [as in (5.45)]

$$\rho_a(t) = \operatorname{sinc}[(1 - a)Ft/2]\operatorname{sinc}[(1 + a)Ft/2] \tag{6.43}$$

To find ρ_b, the transform of $\operatorname{ramp}(f/F)|U(f)|^2$, we see from (6.42) that we require the product of the ramp function with a convolution of two sinc functions. Now in general it is not the case that $u(v \otimes w) = (uv) \otimes w$, but in the particular case where w is a δ-function at the origin, then, as $\delta(x) \otimes y(x) = y(x)$, this relation is true [i.e., $u(v \otimes \delta) = uv = (uv) \otimes \delta$]. In this case, where a is near unity, the smaller rect function [with the factor $2/(1 - a)F$, to make its integral unity] is near a δ-function, and we will make the small approximation of rearranging the product with the convolution in the form

$$\operatorname{ramp}\left(\frac{f}{F}\right)|U(f)|^2 \approx \frac{4}{(1 + a)(1 - a)F^2} \operatorname{rect}\left[\frac{2f}{(1 - a)F}\right] \otimes \operatorname{ramp}\left(\frac{f}{F}\right)\operatorname{rect}\left[\frac{2f}{(1 + a)F}\right] \tag{6.44}$$

Figure 6.12 shows the scale of the approximation. In Figure 6.12(a), the lowest trace (dotted) is the straight line of the falling edge of the trapezoid,

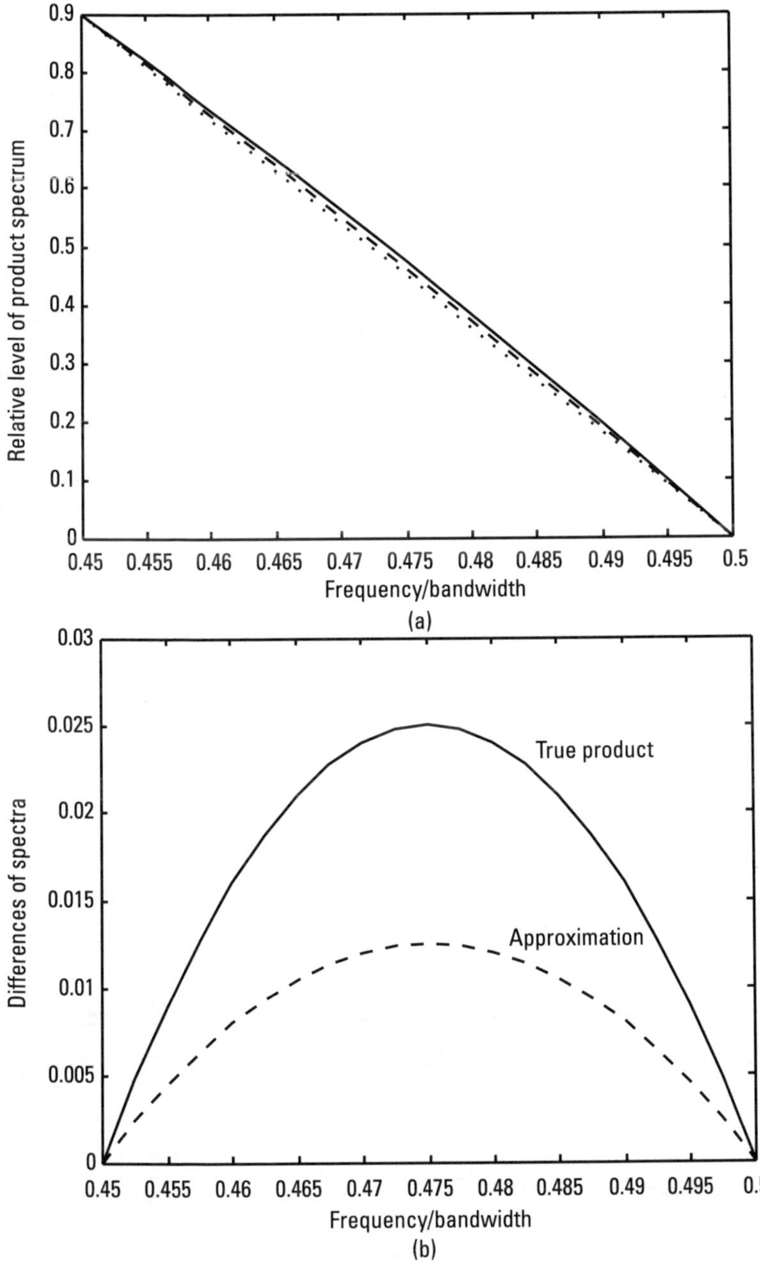

Figure 6.12 Effect of approximation of product spectrum on falling edge of trapezoid: (a) falling edge of product spectra; (b) differences of spectra.

which results if we take a δ-function instead of the factor $[2/(1-a)F]$ rect $[2f/(1-a)F]$. The highest trace (continuous) is a shallow quadratic given by the product of the trapezoid edge with the ramp function over this interval, and is the correct shape. The middle trace (dashed), an even shallower quadratic, corresponds to (6.44), the result of convolving the narrow rect function with the product of the ramp function with the wider rect function, which is illustrated in Figure 6.13. Figure 6.12(b) shows the difference between the two curves and the linear response. There is no error resulting from the rearrangement of (6.44) over the central part of the spectrum, $-aF/2 < f < aF/2$.

We now consider just the product of the ramp function with the wider rect function. As the rect function is narrower than the ramp function, the product is smaller than the unit ramp function, which reaches values of +1 and −1 at its edges, so the result, as illustrated in Figure 6.13, is a scaled ramp function; the scaling factor is the relative width of the rect function, which is $(1+a)/2$. The spectrum to be transformed is thus

$$\operatorname{ramp}\left(\frac{f}{F}\right)|U(f)|^2 = \frac{4}{(1+a)(1-a)F^2}\,\frac{(1+a)}{2}\operatorname{ramp}\left(\frac{2f}{(1+a)F}\right) \otimes \operatorname{rect}\left(\frac{2f}{(1-a)F}\right) \tag{6.45}$$

and its transform is, using (6.18),

$$\rho_b(t) = i\left(\frac{1-a}{2}\right)\operatorname{snc}_1\left[\frac{(1+a)Ft}{2}\right]\operatorname{sinc}\left[\frac{(1-a)Ft}{2}\right] \tag{6.46}$$

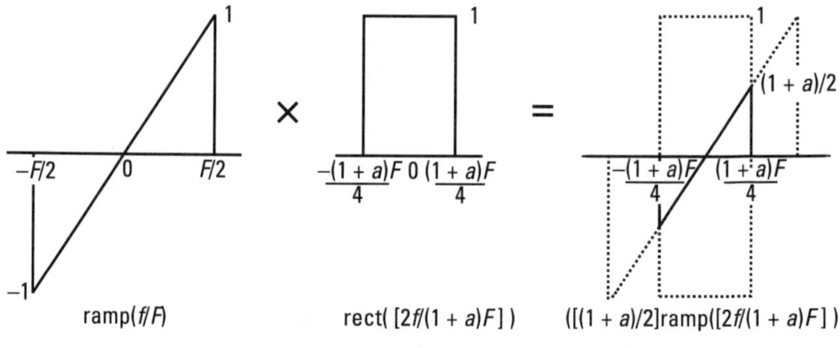

Figure 6.13 Product of ramp and rect functions.

Finally, for ρ_c the function to be transformed is $\text{ramp}^2(f/F)|U(f)|^2$, and (again making the small approximation by rearranging the expression) we can see that the product of $\text{ramp}^2(f/F)$ with $\text{rect}[2f/(1+a)]$ is $[(1+a)/2]^2\,\text{ramp}^2\,[2f/(1+a)]$, and, again using (6.18), the transform is given by

$$\rho_c(t) = -\left(\frac{1+a}{2}\right)^2 \text{snc}_2\left[\frac{(1+a)Ft}{2}\right]\text{sinc}\left[\frac{(1-a)Ft}{2}\right] \quad (6.47)$$

Using (6.43), (6.46), and (6.47) to substitute for ρ_a, ρ_b, and ρ_c in (6.40) and (6.41), and also putting $FT = 1/q$, as the sampling interval is the reciprocal of the (oversampled) sampling rate qF, we obtain

$$a_r = \text{snc}_0(\alpha_1)\left[\text{snc}_0(\alpha_2) - i\frac{(1+a)\phi}{4}\text{snc}_1(\alpha_2)\right] \quad (6.48)$$

and

$$b_{rs} = \text{snc}_0(\beta_1)\left[\text{snc}_0(\beta_2) + i\frac{(1+a)\phi}{2}\text{snc}_1(\beta_2) - \frac{(1+a)^2\phi^2}{16}\text{snc}_2(\beta_2)\right] \quad (6.49)$$

where

$$\alpha_1 = \frac{(1-a)(r-\beta)}{2q},\ \alpha_2 = \frac{(1+a)(r-\beta)}{2q},\ \beta_1 = \frac{(1-a)(r-s)}{2q},\ \text{and}$$

$$\beta_2 = \frac{(1+a)(r-s)}{2q}$$

Using these expressions for the components of **a** and **B**, we compute the weights of the equalization filters for each element and then plot the difference beam patterns in Figure 6.14 corresponding to the sum beam patterns of Figures 6.5 and 6.6(a). In this case, however, we plot the linear response with angle (rather than the logarithmic power response) in order to show the response passing through zero at the required angular position.

As before, we look at the response in the steered direction as a function of frequency; in this case we require the gain in this direction to be zero and the slope to be constant. The variation of gain over the normalized

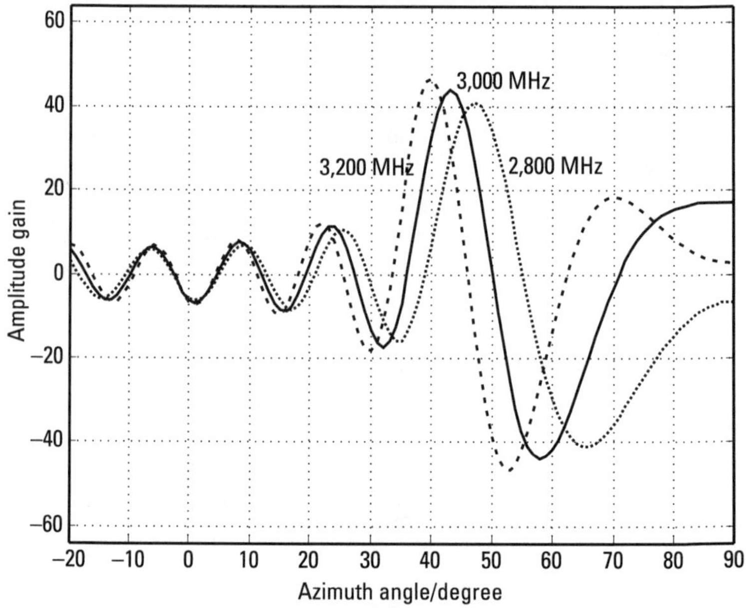

(a)

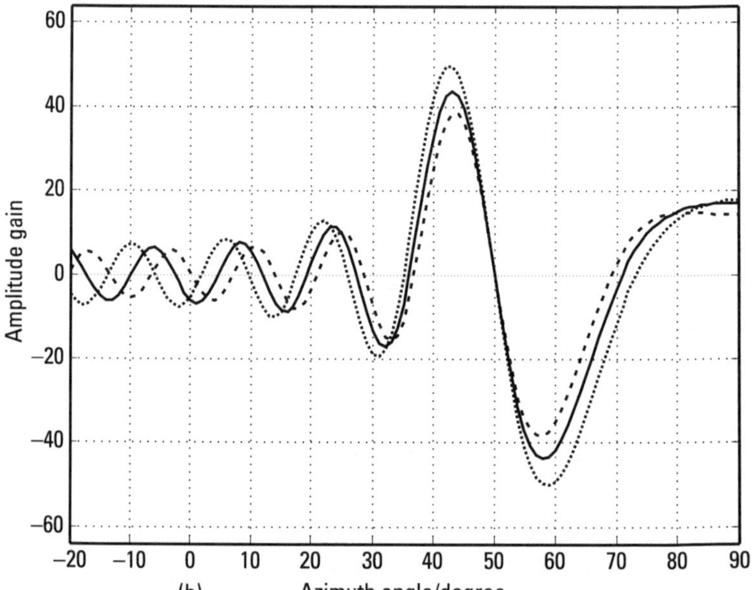

(b)

Figure 6.14 Difference beam response: (a) difference beam with narrowband weights; (b) difference beam with equalization.

bandwidth at baseband is shown in Figure 6.15. (This is for a fractional bandwidth of 100%.) We first show the gain in linear form in Figure 6.15(a). The unequalized response in the look direction, as a function of frequency, is rather similar to the response as a function of direction at the center frequency, shown in Figure 6.14. The power response, in decibels, is shown in Figure 6.15(b), but neither of these shows clearly how well the gain has been kept to zero in the look direction across the band. Changing the scale, in Figure 6.15(c), shows that the gain ripples are some 35 dB below the peaks of the difference beam response, and this is with only five tap filters and oversampling at 20%. Increasing either of these will rapidly reduce the ripple level to much lower values, as shown in Figure 6.15(d), where there are seven taps and 50% oversampling, giving ripples over 20 dB lower. We note that the ripple pattern is not symmetrical about the look direction. In fact, it would be so if we had performed optimum equalization of the difference beam, which only requires delay compensation, rather than its slope. In this case, we have equalized the pattern slope, which requires compensation for both delay and the amplitude variation with frequency seen in (6.35), and as this amplitude rises with frequency the compensation factor (like K in Figure 6.3) falls, and we see that the higher frequency ripples in Figure 6.15(c, d) are indeed smaller than the corresponding ones at lower frequency.

Finally we show, in Figure 6.16, the equalized slope for the two sets of filter parameters used in Figure 6.15. Figure 6.16(a, b) shows the difference between the unequalized and equalized responses. We see that the equalization has been remarkably effective. The equalization using only integral delays gives a considerable improvement, but is still far from adequate. It is slightly better with the higher sampling rate used in Figure 6.16(b). The nearly flat equalized responses are shown amplified in Figure 6.16(c, d). In the first case the variation is a few tenths of a decibel, and with the slightly longer filter and greater sampling rate, it is only a few hundredths (except at the band edges, where the signal power is falling rapidly).

It should be emphasized that Figures 6.15 to 6.17 are for the case of 100% bandwidth—the bandwidth is equal to the center frequency (for example, 100 to 300 MHz). As pointed out following Figure 6.9, the fractional bandwidth is not very significant, except that, of course, as the actual bandwidth increases the sampling rate rises correspondingly, so that while it may be possible to achieve equalization over remarkable fractional bandwidths in principle, in practice there may be difficulty sampling fast enough (and oversampling, while highly desirable, will increase this difficulty). If we consider different bandwidths, we simply see that the initial

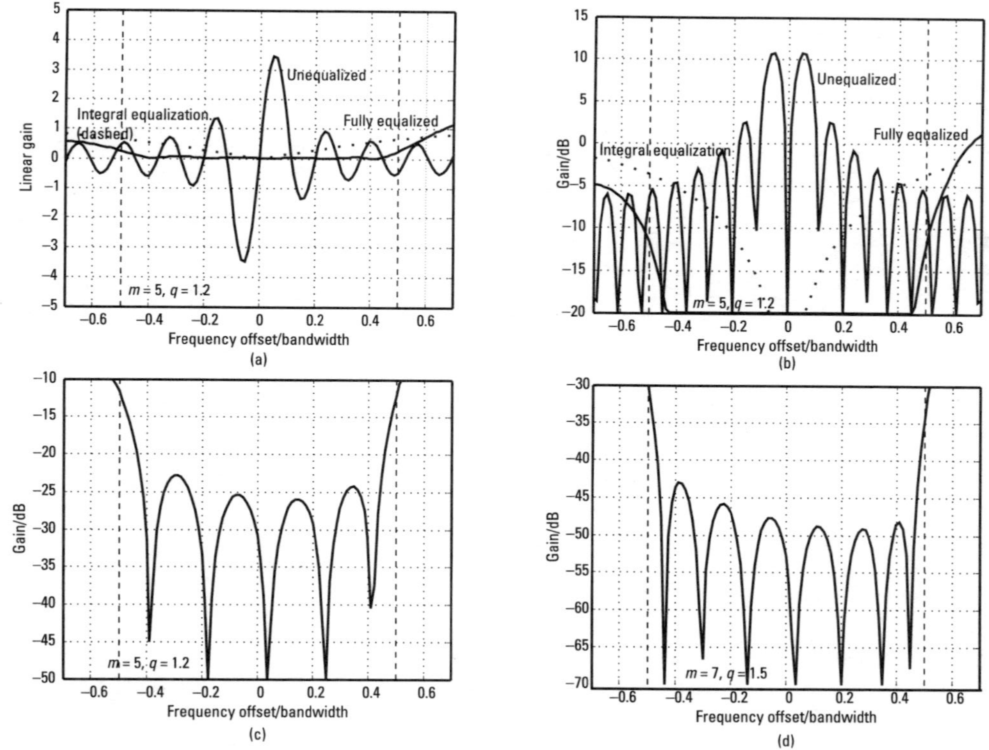

Figure 6.15 Difference beam gain response against frequency offset: (a) linear response; (b) logarithmic response; (c) expanded equalized response; (d) response with more processing.

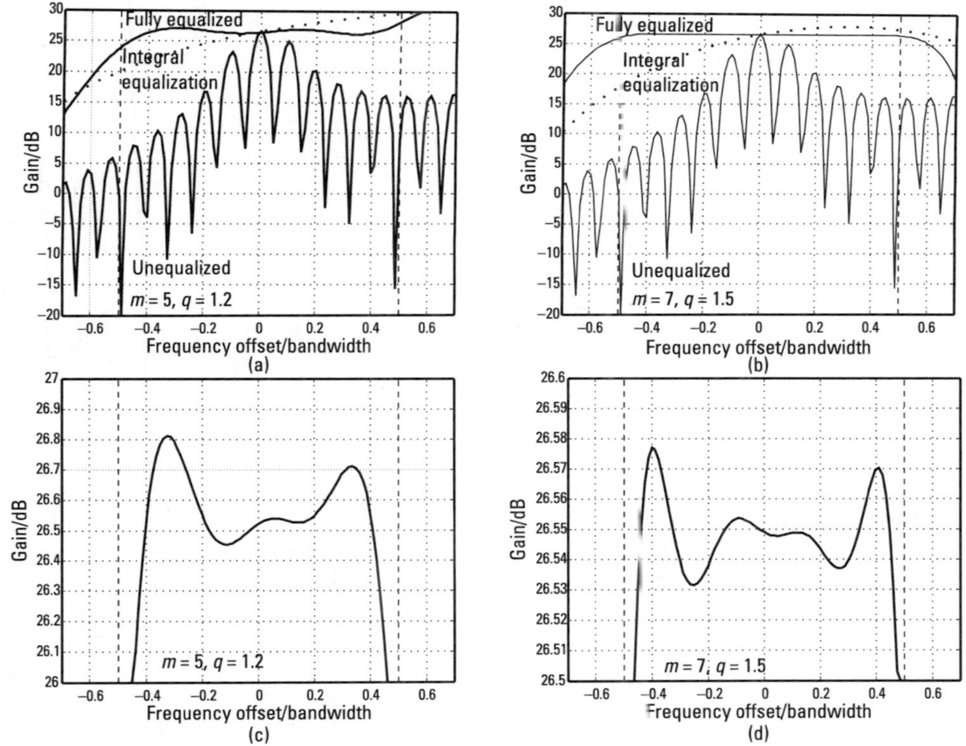

Figure 6.16 Difference beam slope: (a) small filter response; (b) larger filter response; (c) expanded small filter response; (d) expanded larger filter response.

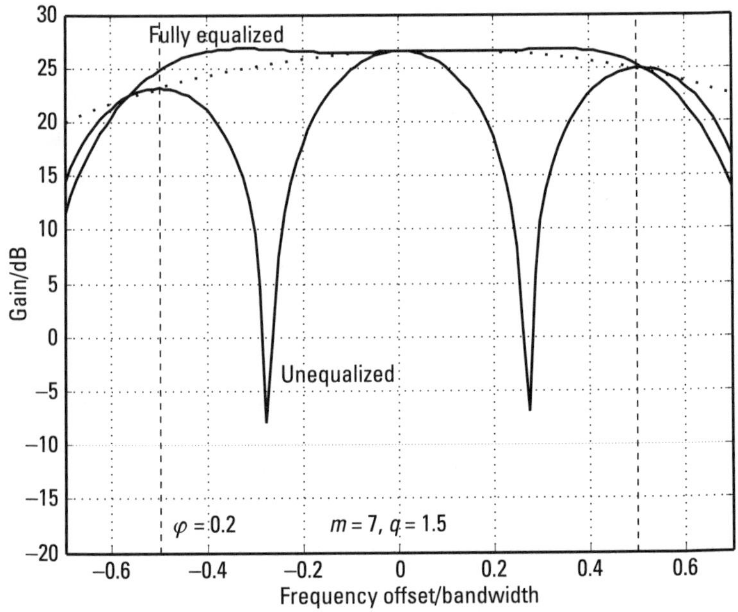

Figure 6.17 Difference beam slope, 20% bandwidth.

equalization problem is different. Figure 6.17 shows the case of a 20% bandwidth (e.g., 180 to 220 MHz). We see that there is less sensitivity (fewer lobes) in the unequalized response, but the equalized result is comparable. If the fractional bandwidth is small enough (e.g., 1% in this case), the unequalized response may be flat enough for equalization not to be necessary, of course, and this is when the narrowband solution is adequate.

6.8 Summary

In this chapter we have looked at equalization of both linear phase variation (due to delay error) and polynomial amplitude error across the band of interest. In the latter case, we saw that the amplitude response requiring equalization could be expressed as a sum of ramp functions. The equalizing weights that minimize the weighted mean square error across the signal band are found as the solution of a matrix equation, the components of which are values of the Fourier transforms of the distorted responses. Thus, for the amplitude distortions, we require the transforms of the ramp functions, and these are found to be derivatives of the sinc function. Including the

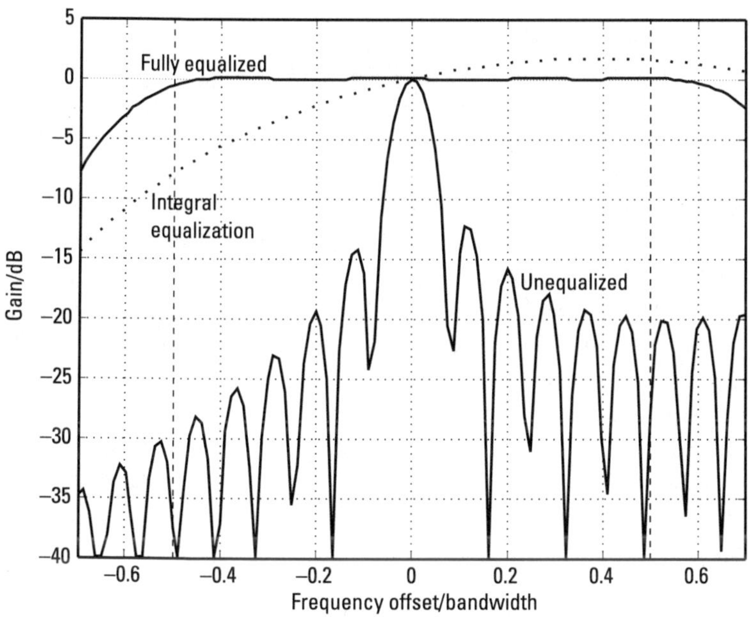

Figure 6.18 Sum beam gain with frequency sensitive elements.

ramp^r – snc$_r$ pairs in the set of transform pairs, we now have the tools for carrying out effective equalization for a range of problems without having to perform explicitly any integration whatever.

After showing that the method is successful in a single channel, including compensating for both amplitude distortion and delay mismatch, the case of forming sum and difference beams using an array was taken. Very effective equalization indeed was found to be possible, and, as shown in the interpolation study of Chapter 5, quite short equalization filters are adequate for high performance if there is some degree of oversampling. This does not mean sampling at several times the minimum rate, but typically at only 20% to 50% higher. Only a simple array, of 16 elements in a regular linear configuration, was taken, but the method is general and is also applicable to much larger arrays and arrays of different configurations, such as regular or irregular, planar or volume arrays. Each complex digital channel (whether fed by a single element or a subarray) has a delay and amplitude response that requires equalization, and the process is the same however the elements are distributed.

In the example, the elements were taken to be frequency independent, so the sum beam equalization required only delay compensation. However, for the difference beam slope it was found that linear amplitude compensation is required as well. If the elements were frequency sensitive, then the equalization could be made to include this effect. We illustrate the case of element amplitude sensitivity (taken to be proportional to frequency at RF) on the sum beam in Figure 6.18. We note higher lobes in the unequalized response at the higher frequencies due to the element responses (and also the similarly asymmetric partially equalized response), but the equalized response is flat to a high degree of accuracy.

Finally we noted that the effectiveness of the equalization is largely independent of actual fractional signal bandwidth (the ratio of the bandwidth to the center frequency), and bandwidths up to 200% (from zero to twice the carrier frequency) can be handled, though of course wider signal bandwidths require proportionally higher sampling rates (further raised by oversampling.)

7

Array Beamforming

7.1 Introduction

In this chapter we consider how the rules and pairs technique can be applied to relate aperture distributions to antenna beam patterns, particularly for antennas made up of linear arrays of similar elements. We start by showing there is a Fourier transform relationship between linear aperture distributions and beam patterns. This relationship holds in general, including for continuous distributions, but we subsequently restrict our study here to multielement arrays, which are in effect discretely sampled apertures. In the case of uniform, or regular (evenly spaced), linear arrays, the aperture distribution is of the form of a comb function, which has a rep function as its transform. For this kind of array, the rules and pairs method works well and is easy to apply for suitable problems; examples are given where, in one case, a simple beam is required (with further study of variations with low side-lobe patterns) and in another case a beam covering a sector at a uniform gain level is generated.

If the array elements are not uniformly distributed, then the convenience of the comb/rep transform is not applicable and a more general least squares error solution is required. As the Fourier transform is a least squares error solution also, this general approach, if applied in the uniform array case, would result in the same solution, if not quite so directly achieved. This approach still requires Fourier transforms and is presented in Section 7.4. Although we cannot use the Fourier transform of the aperture in the general array case, the transform is still required in determining the components of the matrix and vector used to obtain the weights. The Fourier transform is also useful for general results regarding the relationship between the weights and the patterns, as shown in Sections 7.2 and 7.3.

In this chapter we consider only the narrowband case, where the bandwidth is small enough for the effect of delay across the aperture to be adequately approximated as a phase shift at the center frequency of operation. This condition holds for a very wide range of radio and radar problems, but when it does not, the equalization methods of Chapter 6 can be applied. We also consider only the case of the linear aperture as, again, this is very widely encountered, and in the form of the uniform linear array (ULA) is particularly suitable for analysis by the rules-and-pairs method. Furthermore, the linear solution is also applicable to regular rectangular *planar* arrays, for which the two-dimensional beam pattern (in direction cosine coordinates u and v, say) is simply the product of the two patterns given by the orthogonal linear apertures.

7.2 Basic Principles

Given a linear aperture, the far field signal strength is proportional to the sum across the aperture of the current at each point weighted by a phase factor that depends on the position in the aperture and the direction for which the response is to be calculated (Figure 7.1). If the signal phase at the reference point is $\phi(t)$, then at the point x along the aperture it is $\phi(t+\tau)$, where $\tau = x \sin \theta / c$ is the time taken for the wavefront at x to reach the reference point, and c is the speed of light. The phase at point x, at signal frequency f_0, is thus $2\pi f_0 \tau$, or $2\pi x \sin \theta / \lambda_0$, where λ_0 is the corresponding wavelength and $f_0 \lambda_0 = c$.

If the aperture current distribution is $a(x)$, then the relative gain in direction θ, measured from the normal to the array, is given, within a scaling factor, by

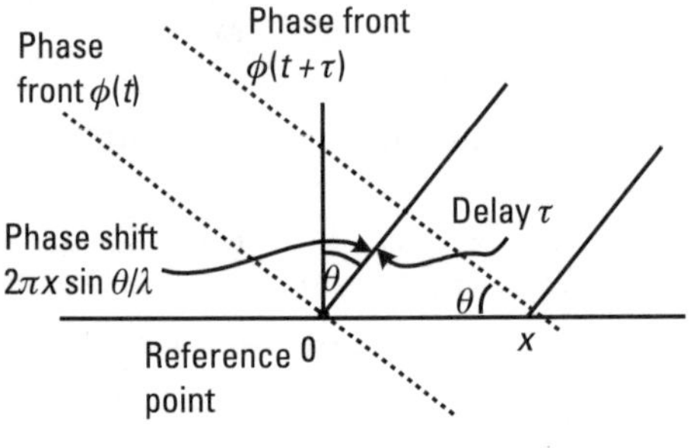

Figure 7.1 Aperture phase shift.

$$g(\theta) = \int_{-\infty}^{\infty} a(x) \exp(2\pi i x \sin\theta/\lambda)\, dx \tag{7.1}$$

The integral is over the whole domain of x, though with a finite aperture $a(x)$ will be zero outside this finite region. If we define the aperture positions in units of the wavelength of operation λ_0, then we replace x/λ_0 with x, and also if we define $u = \sin\theta$, as in Section 6.7, then (7.1) becomes

$$g(u) = \int_{-\infty}^{\infty} a(x) \exp(2\pi i x u)\, dx \tag{7.2}$$

and we see that g is formally the inverse Fourier transform of the aperture distribution a, and correspondingly the distribution a is the Fourier transform of the pattern g. However, we must treat this with some caution, because although (7.2) defines values for $g(u)$ when $|u| > 1$, these u values do not correspond to real directions. If we wanted to determine the aperture distribution for a given pattern, and the pattern is defined only for the real angles $-\pi/2 \le \theta \le \pi/2$, then we only have the information for the integration over this finite interval for u $(-1 \le u \le 1)$. However, if g can be defined as the required function in this range of u but the function extends outside this range, then we can integrate over the whole u domain, knowing that the resultant aperture distribution a will give the required pattern over the basic interval. An example is the case of a uniform aperture distribution $a(x) = \text{rect}(x/X)$, where the aperture is given by $-X/2 \le x \le X/2$ and the distribution is uniform over this interval. This has the transform $g(u) = X \,\text{sinc}\, Xu$, a sinc function response with first zeros at $\pm 1/X$. This response is curtailed, for the pattern over real angles, at $\pm\pi/2$ rad (i.e., for $u = \pm 1$). However, if we were given that the required pattern over the real angles $(-1 \le u \le 1)$ is sinc Xu, by integrating sinc Xu over the whole range of u $(-\infty < u < \infty)$, we obtain the rect function for the aperture distribution, which gives the wanted pattern in the real angle region.

In the case of an array of identical elements, with their patterns aligned in parallel, we can partition, or factorize, the array response into an array factor, which would be given by using omnidirectional elements, and the element factor, which multiplies the array factor at each angle. The array factor is obtained by summing the contributions from each element with the appropriate phase factor, as in (7.1). For an array of elements we have a sampled aperture; we can still use the Fourier transform, but the aperture distribution is now described by a set of delta functions. If the array is taken to be a regular linear array, we note that a regular set of delta functions corresponds to the transform of a periodic function, so we expect the array factor to be periodic in this case. If we do not want the pattern to be periodic in the real angle region, we could make the period such that it has just one cycle in this interval, requiring it to repeat at a period of 2 in u. This will correspond to the element separations being ½ (i.e., half a wavelength), a well-known result for a pattern free from grating lobes, for all steered directions. (It could also have a greater repetition period than 2, but this would require an element separation closer than a half wavelength; however, this is undesirable, increasing mutual coupling and causing driving impedance problems on transmission.) If the main lobe is narrow and is fixed at broadside to the array (at $\theta = 0$), then a repetition period in u of just over 1 could be allowed, corresponding to an element separation of just under one wavelength. (With a period of unity in u, repetitions of the main beam, that is, grating lobes, will occur at $u = \pm 1$, which lie along the line of the array, and also at higher integral values for u, of course, which are not in real angle space.)

Finally, we note that, as $\sin(\pi - \theta) = \sin\theta = u$, if we consider the array factor pattern from $-\pi$ to π rad, or -180 degrees to $+180$ degrees, we see that the pattern from 90 degrees to 180 degrees is the reflection, about 90 degrees, of the pattern from 90 degrees to 0 degrees and similarly on the other side—in other words, the pattern has reflection symmetry about the line of the array. Thus, if a main lobe is produced at angle θ_0 degrees, then there will be an identical lobe at $180 - \theta_0$ degrees, and, in particular, if there is a broadside main beam (at 0 degrees), there will be a lobe of equal size at 180 degrees. Later in this chapter we take the case of reflector-backed elements that have a $2\sin[\pi\cos(\theta/2)]$ pattern for $-\pi/2 \le \theta \le \pi/2$ and a response of zero for $\pi/2 \le |\theta| \le \pi$, and this removes the unwanted response in the back direction.

7.3 Uniform Linear Arrays

7.3.1 Directional Beams

Initially we consider a uniform weighting over the aperture of width X. If the element separation is d wavelengths, then the aperture distribution function is given by

$$a(x) = \text{comb}_d\,[\text{rect}\,(x/X)] \tag{7.3}$$

and the beam pattern is (from P3b, R5, and R8b)

$$g(u) = (X/d)\,\text{rep}_{1/d}\,[\text{sinc}\,(Xu)] \tag{7.4}$$

If we want the beam to be steered in some direction u_1, then we require the pattern shape to be of the form $\text{sinc}\,[X(u - u_1)]$ instead of $\text{sinc}\,(Xu)$; this will place the peak of the sinc function at u_1 rather than at zero. Transforming back to the aperture domain (using R6a), we see that this requires the distribution to be

$$a(x) = \text{comb}_d\,[\text{rect}\,(x/X)\exp\,(-2\pi i u_1 x)] \tag{7.5}$$

We see we need to put an appropriate phase slope across the aperture to steer the beam (i.e., offset it in the angle domain). If, on the other hand, we offset the array in the aperture domain, so that the distribution is given by $a(x) = \text{comb}_d\,\{\text{rect}[(x - x_1)/X]\}$, then (by R6b) the pattern is

$$g(u) = (X/d)\,\text{rep}_{1/d}\,[\text{sinc}\,(Xu)\exp\,(2\pi i u x_1)] \tag{7.6}$$

and there is a phase slope with angle across the pattern. (This will have little significance in practice, as there is normally no reason to combine or compare signals received at different points in the far field except for reasons of pattern synthesis.)

The distinction between the patterns in the u domain and in the real-angle domain is illustrated in Figure 7.2. An array of 16 elements was taken with an element spacing of 2/3 wavelengths, which gives a repetition period for the pattern of 1.5 in u. This is shown (in decibel form) in Figure 7.2(a), and this pattern is described by (7.4), repeating as expected, even though values of u outside the interval [−1, 1] do not correspond to real angles. The vertical lines show the segment of the u pattern that corresponds to real angles. In Figure 7.2(c), the beam has been steered to 60 degrees, and we see that the pattern has moved along so that a second beam, a grating lobe, lies within this interval. Figure 7.2(b, d) shows the corresponding real beams plotted over the full 360-degree interval. These figures show two significant differences—the stretching of the pattern towards the ±90 degrees directions with the lobes becoming wider, and the reflection of the pattern about these directions. If the patterns in u-space and angle space are g_u and g_θ, then the gain in direction θ is given by $g_\theta(\theta) = g_u(\sin\,\theta)$.

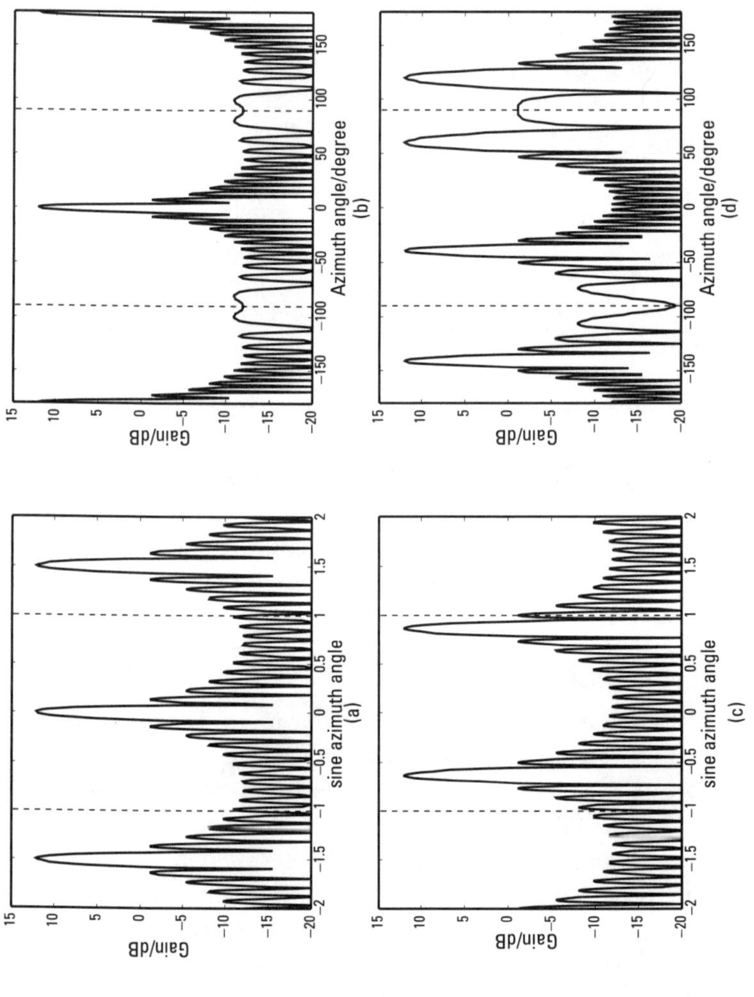

Figure 7.2 Beam patterns for a uniform linear array: (a) broadside beam, *u*-space; (b) broadside beam, angle space; (c) beam at 60 degrees, *u*-space; (d) beam at 60 degrees, angle space.

In plotting this curve, (7.4) was not used, as that would require summing a large number of sinc functions—in principle an infinite number. We can describe the aperture distribution given in (7.3) alternatively by

$$a(x) = \sum_{k=-(n-1/2)}^{(n-1)/2} \delta(x - kd) \tag{7.7}$$

where n is the number of elements in the aperture A [and so $(n - 1)d \leq A < nd$].

This has the (inverse) transform, from R6b,

$$g(u) = \sum_{k=-(n-1/2)}^{(n-1)/2} \exp(2\pi ikdu) \tag{7.8}$$

and this finite sum is much easier to evaluate. However, the form given in (7.4) is still useful, making much more explicit the periodic form of the pattern in the u domain.

7.3.2 Low Sidelobe Patterns

In Sections 3.4 to 3.7, the spectrum of a pulse was shown to improve, in the sense of producing lower sidelobes and concentrating the spectral energy in the main lobe, by reducing the discontinuities (in amplitude and slope) at the edges of the pulse. The same principle is applicable for improving antenna patterns by shaping (or *weighting, tapering,* or *shading*) the aperture distribution in the same way—in fact, if the aperture distributions are given by the pulse shapes of Chapter 3, the beam patterns (in u-space) will be the same as the pulse spectra, as the same Fourier relationship holds. (Strictly speaking, for the pulse spectra the forward Fourier transform is required, while for the beam patterns it is the inverse transform. However, for the frequently encountered case of symmetric distribution functions, there is no distinction.) This is actually the case for continuous apertures, but in the case of a regular linear array, corresponding to a sampled aperture, the pattern is repetitive and is given (over the fundamental interval $-1 \leq u \leq 1$) by the sum of repeated versions of the continuous aperture pattern [as in (7.4) and (7.6) for the rectangular distribution]. For a reasonably narrow beam, particularly one with low sidelobes, the effects of the overlaps will be very small and often negligible. Figure 7.3 shows array patterns for a regular linear array, again of 16 elements, for both the unweighted case (rectangular aperture weighting, dotted line) and raised cosine weighting (solid line).

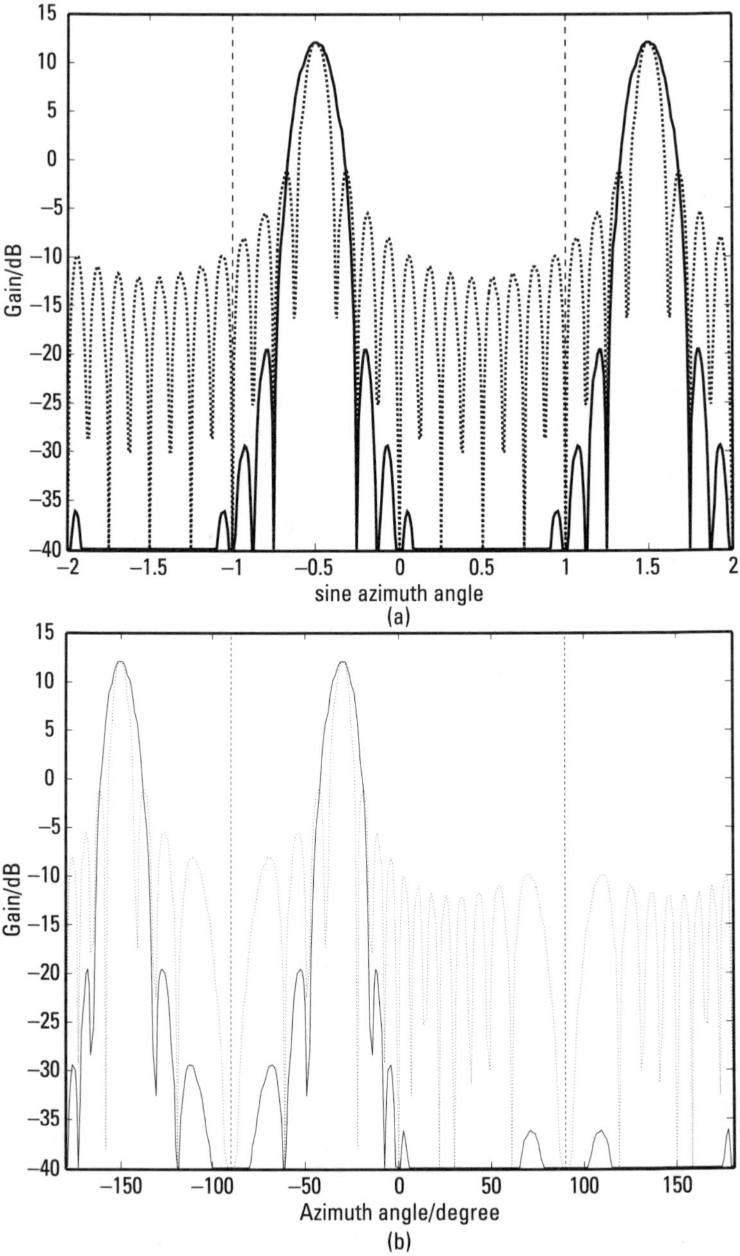

Figure 7.3 Beam patterns for uniform linear array with raised cosine shading: (a) *u*-space; (b) angle space.

In this case, the aperture distribution is given by $[1 + \cos(2\pi x/X)]$ with transform (as in Section 3.4, with $X = 1/U$ replacing $2T = 1/f_0$, and omitting the scaling factor) $\operatorname{sinc}(u/U) + ½ \operatorname{sinc}[(u - U)/U] + ½ \operatorname{sinc}[(u + U)/U]$. The figure shows both the response in u-space and with angle, as in Figure 7.2, but in this case the element spacing is 0.5 wavelength [so the repetition interval in u is 2, as seen in Figure 7.3(a)] and the beam direction is −30 degrees. The weighting has been very effective in reducing the side-lobe levels, though at the cost of broadening the main lobe.

Clearly we could apply different weighting functions, obtaining the corresponding beam patterns, given by their Fourier transforms, but this would be simply going over the ground of Chapter 3, where pulses of various shapes and their spectra were studied. Instead, we look at two other possibilities for improving the pattern, not necessarily for practical application, but as illustrations of approaches to problems of this kind that could be of interest. First, we note that the main lobe in Figure 7.3 consists of the sum of the main sinc function with two half amplitude sinc functions, offset on each side by one natural beam width (the reciprocal aperture; this is actually the beam width at 4 dB below the peak). This suggests continuing to use sinc functions to obtain further improvement. We could reduce the largest side lobes, near ±2.5 beamwidth intervals by placing sinc functions of opposite sign at these positions. This will have to be done quite accurately, because these side lobes are already at about −31 dB below the peak, or at a relative amplitude of 0.028, so an amplitude error of 1%, for example, would not give much improvement. To find the position of these peaks, we can use Newton's method for obtaining the zeros of a function. In this case, the function is the *slope* of the pattern, as we want the position of a lobe rather than a null. In this discussion, we neglect the overlapping of the repeated functions on the basis that, for an aperture of moderate size (such as that of this 16-element array, which is effectively eight wavelengths) the effect of overlap is small, especially in the low-side-lobe case—in fact, by dropping the rep function, we are studying the pattern of the continuous aperture. In addition, we plot the pattern in units of the beam width U, as this simply acts as a scaling factor (in u-space).

Differentiating the expression above for the beam shape $g(u)$ to obtain its slope $g'(u)$, we have

$$g'(u) = (u/U)\left\{\operatorname{snc}_1(u/U) + \frac{1}{2}\operatorname{snc}_1[(u - U)/U] + \frac{1}{2}\operatorname{snc}_1[(u + U)/U]\right\} \tag{7.9}$$

where $\pi\,\mathrm{snc}_1$ is the derivative of the sinc function, as defined in Section 6.3 above [see (6.17)]. Using Newton's approximation method to find the peak of a lobe (a point of zero slope), we have

$$u_{r+1} = u_r - g'(u_r)/g''(u_r) \tag{7.10}$$

and if we put $v = u/U$, to give the pattern in terms of natural beam widths, then this becomes

$$v_{r+1} = v_r - (1/U)g'(Uv_r)/g''(Uv_r) \tag{7.11}$$

Here u_r and v_r are the approximations after r iterations. Putting in g' from (7.9) and g'' from another differentiation of (7.9), we obtain

$$v_{r+1} = v_r - \frac{2\,\mathrm{snc}_1(v_r) + \mathrm{snc}_1(v_r - 1) + \mathrm{snc}_1(v_r + 1)}{\pi(2\,\mathrm{snc}_2(v_r) + \mathrm{snc}_2(v_r - 1) + \mathrm{snc}_2(v_r + 1))} \tag{7.12}$$

Starting with $v_0 = 2.5$, this converges rapidly (v_4 is equal to v_3 to four decimal places) to give a value of –0.0267 at $v = 2.3619$. Adding sinc functions to cancel the lobes at ±2.5, the pattern in v is now

$$g(v) = \mathrm{sinc}(v) + \frac{1}{2}[\mathrm{sinc}(v-1) + \mathrm{sinc}(v+1)] \tag{7.13}$$
$$+ 0.0267[\mathrm{sinc}(v - 2.362) + \mathrm{sinc}(v + 2.362)]$$

This pattern is shown in Figure 7.4, with the raised cosine shaded pattern for comparison (dotted curve). We see that the original first side lobes have been removed and the new largest side lobes are at almost –40 dB, an improvement of nearly 10 dB. To find the weighting function that gives this pattern, we require the Fourier transform of (7.13). This can be seen almost by inspection by following in reverse direction the route that gave the raised cosine transform. More formally, we have

$$g(v) = \mathrm{sinc}(v) \otimes \Big\{\delta(v) + \frac{1}{2}[\delta(v-1) + \delta(v+1)] \tag{7.14}$$
$$+ 0.0267[\delta(v - 2.362) + \delta(v + 2.362)]\Big\}$$

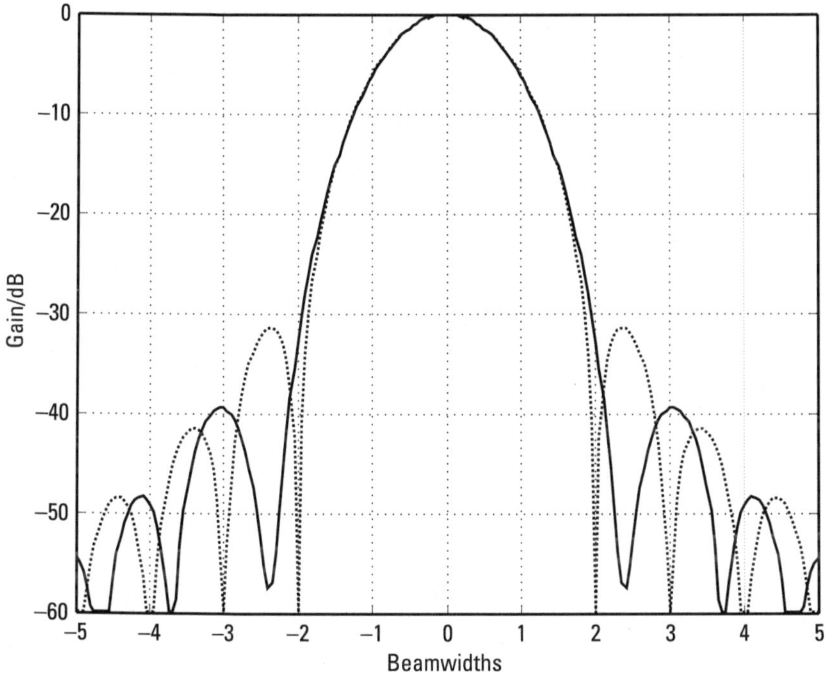

Figure 7.4 Beam pattern for ULA with additional shading.

giving, on Fourier transforming,

$$a(y) = \text{rect}(y)\left\{1 + \frac{1}{2}[\exp(2\pi iy) + \exp(-2\pi iy)] + 0.0267[\exp(2\pi i 2.362y) + \exp(-2\pi i 2.362y)]\right\}$$

$$= \text{rect}(y)[1 + \cos(2\pi y) + 0.0534\cos(4.724\pi y)] \qquad (7.15)$$

As we started with the normalized variable $v = u/U$, this distribution is in terms of the normalized aperture $y = x/X$.

For the second example, we produce a pattern with the closest side lobes to the main beam (and the largest), all at almost the same level, similar to the pattern given by Taylor weighting. In this case, we take the pattern to be given by a sum of sinc functions at $0, \pm 1, \pm 2, \ldots, \pm n$ natural beam widths (reciprocal aperture units) from the center. In this case, we do not take the amplitudes of the sinc functions at ±1 to be 0.5, as above. Thus we have, again using a normalized u-space variable,

$$
\begin{aligned}
g(v) = {} & \operatorname{sinc}(v) + a_1 [\operatorname{sinc}(v - 1) + \operatorname{sinc}(v + 1)] \\
& + a_2 [\operatorname{sinc}(v - 2) + \operatorname{sinc}(v + 2)] \\
& + a_n [\operatorname{sinc}(v - n) + \operatorname{sinc}(v + n)]
\end{aligned}
\tag{7.16}
$$

The n coefficients are determined by setting the gain to particular values at n points in the form $g(v_r) = g_r$. The values we choose are the constant level A, or $-A$, at the side-lobe peaks, where $20 \log_{10}(A)$ is the required peak level in decibels. We do not know exactly where these peaks are, but we should be near the peak positions if we choose the points to be midway between the nulls in the sinc patterns; thus we have

$$
g(r + 1.5) = (-1)^{r+1} A \qquad (r = 1 \text{ to } n) \tag{7.17}
$$

[The factor $(-1)^{r+1}$ is required, as the amplitudes of the side-lobe peak magnitudes alternate in sign.] The set of n equations given by putting the conditions of (7.17) into (7.16) leads to the vector equation $\mathbf{Ba} = \mathbf{b}$, with solution

$$
\mathbf{a} = \mathbf{B}^{-1}\mathbf{b} \tag{7.18}
$$

where $\mathbf{a}$ contains the required coefficients, and the components of $\mathbf{b}$ are given by

$$
b_j = (-1)^{j+1} A - \operatorname{sinc}(j + 1.5) \tag{7.19}
$$

and the components of $\mathbf{B}$ by

$$
B_{jk} = \operatorname{sinc}(j - k + 1.5) + \operatorname{sinc}(j + k + 1.5) \tag{7.20}
$$

We note that the first points, at ±2.5, are on the edge of the main lobe rather than the peak of a separate side lobe, and the value is positive, with subsequent points on (or near) side-lobe peaks and alternating in sign. Two patterns given by (7.16), with coefficients from (7.18), are shown in Figure 7.5, again with the raised cosine pattern for comparison. In Figure 7.5(a), we took $n = 3$ and the required level to be −50 dB. The two separate

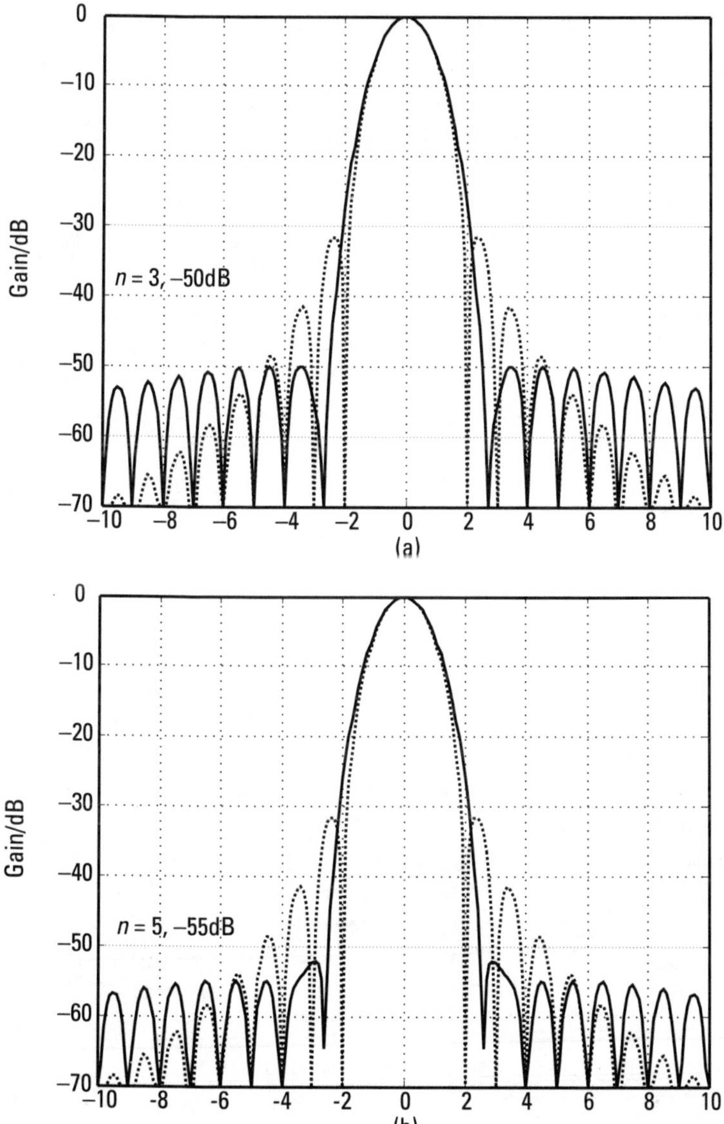

Figure 7.5 Constant-level side-lobe patterns: (a) first two lobes nominally at –50 dB; and (b) first four lobes nominally at –55 dB.

lobes are seen to be very close to this level—the pattern levels at ±2.5, ±3.5, and ±4.5 are precisely −50 dB by construction, but the peaks of the lobes will not be at exactly these points, so the actual peaks will rise slightly above the required value. However, the range of levels for which this works well is limited, and Figure 7.5(b) shows it starting to fail. In this case, $n = 5$ and the nominal level is −55 dB. This is seen to be attained very closely for the lobes at ±4.5, ±5.5, and ±6.5, but the pattern has bulged between ±2.5 and ±3.5, giving a lobe appreciably above the specified level. Nevertheless, these are good side-lobe levels and have been obtained quite easily. The pattern is well behaved when designed for −50-dB sidelobes, but the first sidelobe, near ±2.5, starts to rise when the specified level is about −48 dB or higher. In general, for these patterns the coefficient a_1 is near 0.5 and the other coefficients fall rapidly in magnitude. To find the corresponding weighting function, we transform the pattern to obtain

$$a(y) = \text{rect}(y)\,[1 + 2a_1 \cos(2\pi y) + 2a_2 \cos(4\pi y) + \ldots + 2a_n \cos(2n\pi y)] \tag{7.21}$$

7.3.3 Sector Beams

We now consider a quite different problem—that of providing a flat, or constant gain, beam for reception or transmission over a sector, generally wide compared with the natural beam width. In this case, as we want the sector gain to be constant over an interval (for simplicity, we take the amplitude to be unity), it will be of the form $\text{rect}(u/u_0)$, where the width of the sector is u_0, centered on broadside initially. As we want a regularly sampled aperture distribution for a uniform linear array rather than a continuous one, we take the required pattern to be repetitive in the u domain, to be given by

$$g(u) = \text{rep}_U\,[\text{rect}(u/u_0)] \tag{7.22}$$

so the element weights across the aperture are given by

$$a(x) = (u_0/U)\,\text{comb}_{1/U}\,\text{sinc}(u_0 x) \tag{7.23}$$

This is a sinc function envelope, with width proportional to $1/u_0$ and sampled at intervals $1/U$ wavelengths, where U is the repetition interval in the u domain. If we take the beam to have an angular width θ_0, then the edges of the beam are at $\pm\theta_0/2$ and the corresponding u_0 value is given by

$$u_0 = 2 \sin(\theta_0/2) \tag{7.24}$$

[It is important not to put $u_0 = \sin(\theta_0)$ because of the nonlinear relationship between the variables. For example, if we chose $\theta_0 = 90°$, then the first, correct, expression makes $u_0 = \sqrt{2}$, while the second makes $u_0 = 1$; this would actually give a 60-degree beam.]

Figure 7.6(a) shows an example of a sector beam generated this way, with the weights applied to the elements shown in Figure 7.6(b). The aperture distribution is a sampled sinc function and, for perfect patterns, extends, in principle, over the whole x-axis. In practice, it is limited to n elements so is effectively gated by a rect function, rect(x/nd), where $d = 1/U$ is the separation between elements and nd is the effective aperture. In this case, $U = 2$ and d is a half wavelength. The transform of this rect function is a

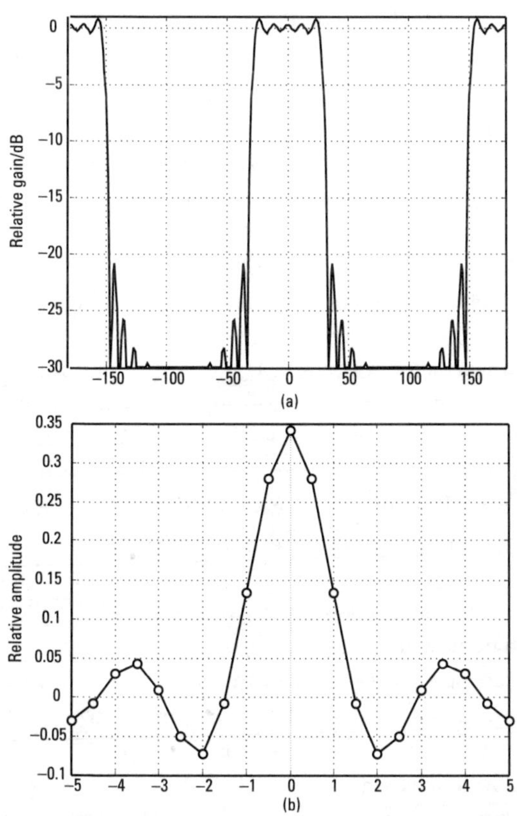

Figure 7.6 Sixty-degree sector beam from an array of 21 elements: (a) beam pattern; (b) element weights.

relatively narrow sinc function; this is convolved with the ideal rectangular pattern given by an infinite array to produce the ripple seen in the figures. The figure is for a nominal 60-degree sector beam (from –30 to +30 degrees) for an array of 21 elements.

The side-lobe ripples indicate the width of the natural beam from this aperture—the main-lobe width between the first zeros would be the width of two of these side lobes. With an even number of elements, the distribution is rather different in appearance, with two equal values in the center, but a very similar beam pattern. There is no simple relation between the side-lobe levels and the number of elements (or whether this number is odd or even)—the levels vary with both the number of elements and the beam width. Because of the repetitive form of the response in the u domain, these lobes are the result of summing the convolution ripples of mainly two basic patterns, as given by the continuous aperture, at a separation of $U = 1/d$, and these may sometimes reinforce and sometimes tend to cancel. The fluctuations (with parameter variation) of these lobes will tend to be greater as the sector width increases and the edges of the beam and its repetition become closer.

We also note the appearance of the back lobe in Figure 7.6(a). In many applications, this is undesirable, whether on transmission, when only half the power goes into the forward lobe, or reception, when interference or external field noise will enter through this lobe. This lobe can be removed by mounting a reflecting plane at a quarter wavelength behind the array (Figure 7.7). Combining the direct signal with the reflected one, effectively arriving at a point a half wavelength back, and including the phase change of π on reflection at a denser medium, the element response becomes $2 \sin[\pi \cos(\theta)/2]$ for a signal at angle θ to broadside. This is in the forward half plane, with (for an infinite plane) no response in the back half plane. This is a pattern with a single broad lobe (Figure 7.7), falling to –3 dB at ±60 degrees, and it increases the directivity of the elements by 6 dB; 3 dB is due to limiting the power to one side of the array, and 3 dB to reducing the beam from a 180-degree semicircle to this 120-degree lobe. Because this response is so flat, it will make very little difference to the shape of sector beams centered at, or near, broadside, though it will more noticeably distort beams steered towards the edges of the forward sector.

If we want to steer the beam so that its center is at θ_1 and its width is still θ_0, then its edges are at $\theta_a = \theta_1 - \theta_0/2$ and $\theta_b = \theta_1 + \theta_0/2$, and the corresponding u values are $u_a = \sin\theta_a$ and $u_b = \sin\theta_b$. In this case, the center of the beam in u space is at $u_1 = (u_b + u_a)/2$ and its width is $u_0 = u_b - u_a$. With these definitions of u_0 and u_1, the required sector beam pattern is, from (7.22),

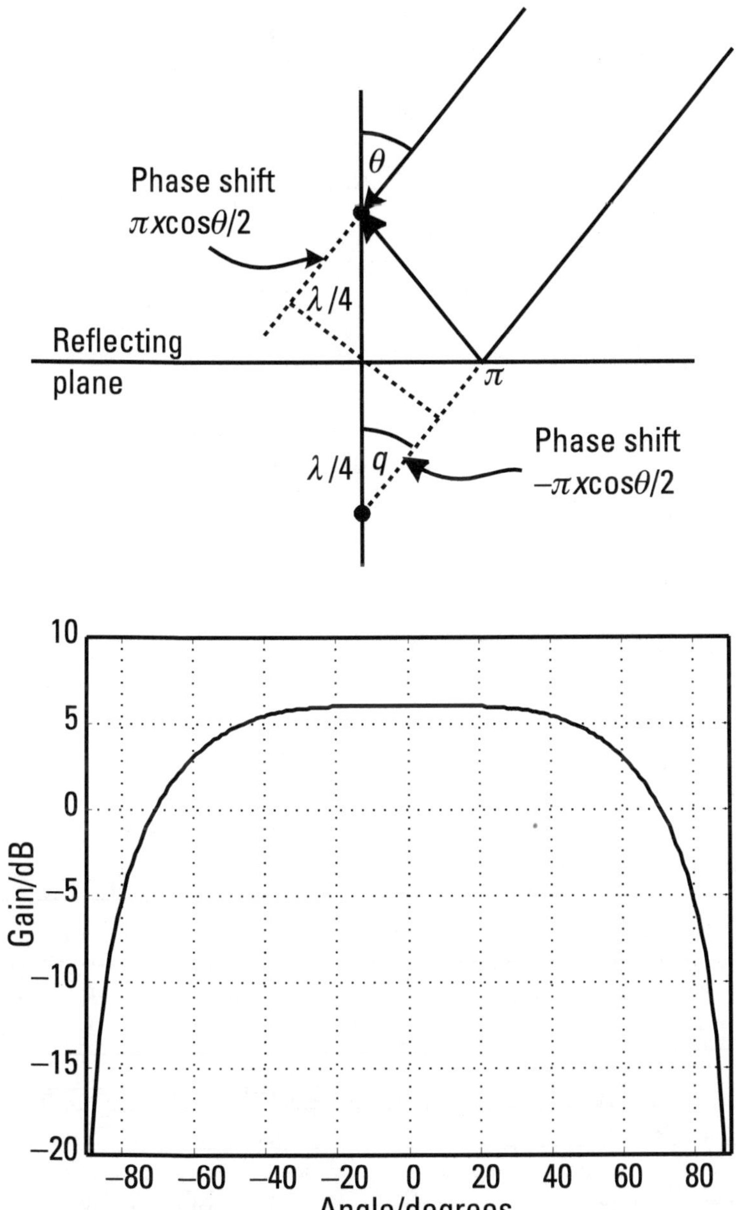

Figure 7.7 Element response with reflector.

$$g(u) = \mathrm{rep}_U\,[\mathrm{rect}\,(u - u_1)/u_0] \tag{7.25}$$

This has the (forward) transform (using R6a)

$$a(x) = (u_0/U)\,\mathrm{comb}_{1/U}\,[\mathrm{sinc}\,(u_0 x)\exp(-2\pi i x u_1)] \tag{7.26}$$

and we see that this requires putting a linear phase slope across the array elements; this corresponds to the effect of the delay across the aperture for a waveform received from (or transmitted to) this direction, causing a phase shift at the carrier frequency f_0. This requires an infinite aperture; for a finite aperture, we put (7.26) in the alternative form of a finite sum of δ-functions, as in (7.7) [but weighted by $(u_0/U)\,\mathrm{sinc}\,(u_0 kd)\exp(-2\pi ikdu_1)$ in this case, with $d = 1/U$], and carrying out the inverse transform gives

$$g(u) = \frac{u_0}{U}\sum_{k=-(n-1/2)}^{(n-1)/2} \mathrm{sinc}\,(u_0 kd)\exp[2\pi i(u - u_1)kd] \tag{7.27}$$

This form of the beam pattern is the alternative, for practical evaluation, to (7.25) with the sinc function convolution due to the finite aperture width incorporated.

Figure 7.8 illustrates a steered sector beam with a reflector-backed array. In this case, the beam is formed from a uniform linear array of 12 elements, at half-wavelength spacing, and is 90 degrees wide, centered at 20 degrees. The response with omnidirectional elements is shown (dotted line) for comparison. The reflector removes the back lobe and also distorts the sector beam slightly. The weights are complex, as indicated in (7.26), and as the pattern is specified to be real, the weight distribution, as the transform, has conjugate symmetry, with the real part symmetric and the imaginary part antisymmetric.

The sector beams defined so far have the same phase across the sector, so that, when used for transmission, the signal received in the far field will have the same phase at points in all directions at the same distance from the center of the array. If we put a phase slope across the pattern, this will not change the power transmitted in a given direction, but will change the weights required. In this case, let the slope be such as to produce a phase difference of r cycles across a unit range of u, where the phase variation is linear in u space. The required pattern, from (7.25), is now

$$g(u) = \mathrm{rep}_U\,\{\mathrm{rect}\,[(u - u_1)/u_0]\exp(2\pi iru)\} \tag{7.28}$$

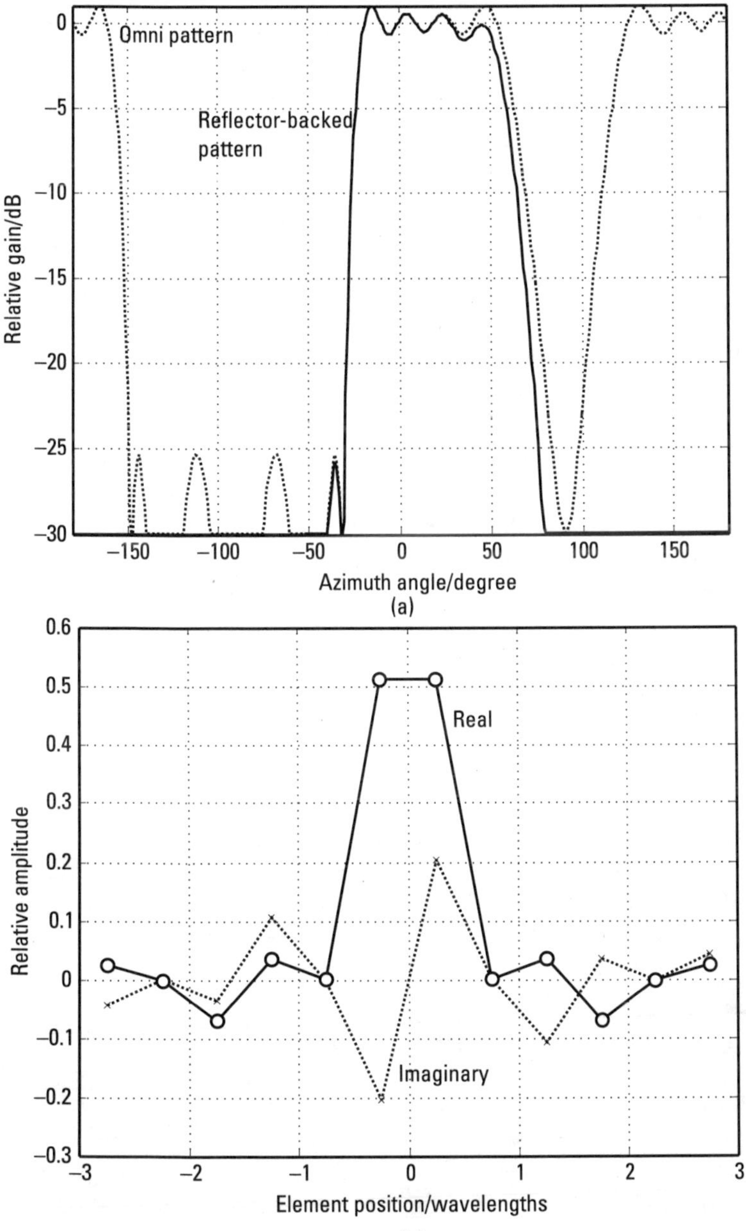

Figure 7.8 Steered sector beam with reflector-backed elements: (a) beam patterns; (b) element weights.

and the weight function, given by the Fourier transform of (7.28), is

$$a(x) = (u_0/U)\,\mathrm{comb}_{1/U}\,[\mathrm{sinc}\,(u_0x)\exp(-2\pi ixu_1) \otimes \delta(x-r)] \quad (7.29)$$
$$= (u_0/U)\,\mathrm{comb}_{1/U}\,\{\mathrm{sinc}\,[u_0(x-r)]\exp[-2\pi i(x-r)u_1]\}$$

We see that the envelope of the set of δ-functions from the comb function, which defines the weights on the elements, is shifted by r wavelengths with this linear phase slope.

Figure 7.9(a) shows the array factor for a 60-degree sector beam from an array of 20 elements at half wavelength spacing, steered to broadside. The beam also has a phase slope of one cycle per unit of u (i.e., $r = 1$), and this requires the sampled sinc function distribution for the weights to be displaced one wavelength from the center of the array, as seen in Figure 7.9(b). As $u = \pm\frac{1}{2}$ at ± 30 degrees, the phase variation should be 360 degrees across this interval, and this is seen in Figure 7.9(c), which shows the phase relative to that at the center of the beam. The slope varies slightly (because of the finite aperture effect, which causes the amplitude ripples, and also due to the stretching of the pattern in angle space, at higher angle values, compared with u space), but is close to the set value.

By splitting the sector into two or more subsectors with linear phase slopes, it is possible to generate beams with more uniform magnitudes, which is desirable for transmitting arrays, with similar power amplifiers on each element. In fact, for a broadside beam from a four-element array, the amplitudes can be made identical (though, of course, there is phase variation between elements) using two subsectors. However, there is some beam shape degradation and for larger arrays there is more difficulty in balancing flatness of the weight magnitudes and quality of the sector beam.

7.4 Nonuniform Linear Arrays

We have seen that the beam pattern in u space is the (inverse) Fourier transform of the aperture distribution, and we can use the rules and pairs technique for a useful range of distributions, for continuous apertures, and (as demonstrated in Section 7.3) for regularly sampled apertures, corresponding to uniform linear arrays. In this case, the regularly sampled aperture is represented as a comb function, for which the transform is known (Rule 8b). However, for nonuniform sampling, no general rule is available and a different approach is required. In this case, given a desired beam shape and a set of element positions, the problem tackled is to find the weights to be

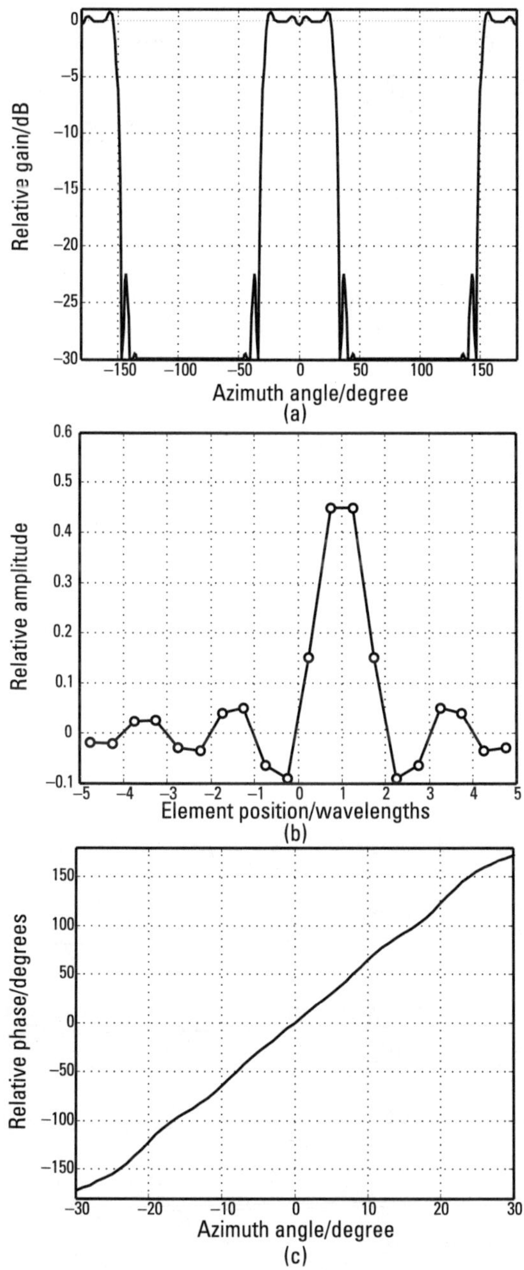

Figure 7.9 Sector beam with phase variation across beam. (a) beam pattern; (b) element weights; and (c) relative phase.

applied to each element to match the desired pattern in a least squared error sense. This problem is very similar to those of Sections 5.3 and 6.2.

For a linear array the aperture distribution is of the form

$$a(x) = \sum_{r=1}^{n} a_r \delta(x - x_r)$$

where the n elements are at positions x_r, with weights a_r. The gain pattern (in u space) is given by the transform of $a(x)$:

$$g(u) = \sum_{r=1}^{n} a_r \exp(2\pi i x_r u) \tag{7.30}$$

Now let $g(u)$ be a desired beam pattern, not necessarily exactly realizable by any linear combination of the n complex exponentials in (7.30). The question is, now, what is the set of n coefficients a_r that gives a least squared error fit to $g(u)$? Let the error at point u be $e(u)$, and defining $f_r(u) = \exp(-2\pi i x_r u)$, we have

$$\begin{aligned} e(u) &= g(u) - \sum_{r=1}^{n} a_r \exp(2\pi i x_r u) \\ &= g(u) - \sum_{r=1}^{n} a_r f_r{}^*(u) \\ &= g(u) - \mathbf{f}^{\mathrm{H}}(u)\mathbf{a} \end{aligned} \tag{7.31}$$

where $\mathbf{a}$ and $\mathbf{f}$ are n-vectors with components a_r and f_r (and the raised suffix H indicates complex conjugate transpose). The square modulus of e is

$$|e(u)|^2 = |g(u)|^2 - \mathbf{f}^{\mathrm{H}}(u)\mathbf{a}g(u)^* - g(u)\mathbf{a}^{\mathrm{H}}\mathbf{f}(u) + \mathbf{a}^{\mathrm{H}}\mathbf{f}(u)\mathbf{f}(u)^{\mathrm{H}}\mathbf{a} \tag{7.32}$$

The total squared error as a function of the weights $\epsilon(\mathbf{a})$ is given by the integral of $|e(u)|^2$ over the interval I in u over which we want the specified response. In some cases this will be the whole real angle region, from $u = -1$ to $u = +1$, but this need not necessarily be the case. The integrated error as a function of the vector $\mathbf{a}$ is thus $\epsilon(\mathbf{a})$, given by

$$\epsilon(\mathbf{a}) = \int_{\mathrm{I}} |e(u)|^2 \, du = p - \mathbf{b}^{\mathrm{H}}\mathbf{a} - \mathbf{a}^{\mathrm{H}}\mathbf{b} + \mathbf{a}^{\mathrm{H}}\mathbf{B}\mathbf{a} \tag{7.33}$$

where $p = \int_{\mathrm{I}} |g(u)|^2 \, du$, and the components of **b** and **B** are given by

$$b_r = \int_{\mathrm{I}} f_r(u) g(u) \, du = \int_{\mathrm{I}} \exp(-2\pi i x_r u) g(u) \, du \tag{7.34}$$

$$B_{rs} = \int_{\mathrm{I}} f_r(u) \, f_s(u) \, du = \int_{\mathrm{I}} \exp[-2\pi i (x_r - x_s) u] \, du$$

The value of **a** that minimizes ϵ (or, more generally, gives a stationary point of ϵ), $\mathbf{a}_0$, is given by $\partial\epsilon/\partial\mathbf{a}^* = \mathbf{0}$ or, from (7.33), $-\mathbf{b} + \mathbf{B}\mathbf{a}_0 = \mathbf{0}$, so that

$$\mathbf{a}_0 = \mathbf{B}^{-1}\mathbf{b} \tag{7.35}$$

This gives the set of weights for the functions $\{f_r\}$ that gives the best fit in a least squares sense to the function g, over the interval I, where the components of **b** and **B** are given in (7.34).

Taking first the case of forming a sector beam from a regular array, let the element separation be d, and so $U = 1/d$ is the pattern repetition interval in the u domain. Then it is natural to choose I to be the interval $[-U/2, U/2]$, which is equivalent to including the factor $\mathrm{rect}(u/U)$ in the integrands in (7.34). In this case, B_{rs} is the Fourier transform of $\mathrm{rect}(u/U)$ evaluated at $(x_r - x_s)$, that is, $U \,\mathrm{sinc}[(x_r - x_s)U]$. However, as $x_r - x_s$ is an integer times d and $dU = 1$, we see that $B_{rs} = U\delta_{rs}$ and $\mathbf{B} = U\mathbf{I}$. For the sector beam, width u_0, centered at u_1, $g(u) = \mathrm{rect}[(u - u_1)/u_0]$, and as this is taken to be within $\mathrm{rect}(u/U)$, the product is still $g(u)$. Then b_r is the Fourier transform of $g(u)$ evaluated at x_r, and we find that the weights a_r given by (7.34) and (7.35) in this case are exactly the same as given by (7.26), rather more directly, confirming the point that the Fourier transform solution is also the least squared error solution.

The solution given by (7.34) and (7.35) is more general than that of (7.23) for the regular array, so that a solution can be found for the weights of an irregular linear array giving a close approximation to a required pattern. Figure 7.10 shows a sector pattern obtained from an irregular array. For this plot, the array elements were displaced from their regular positions, with

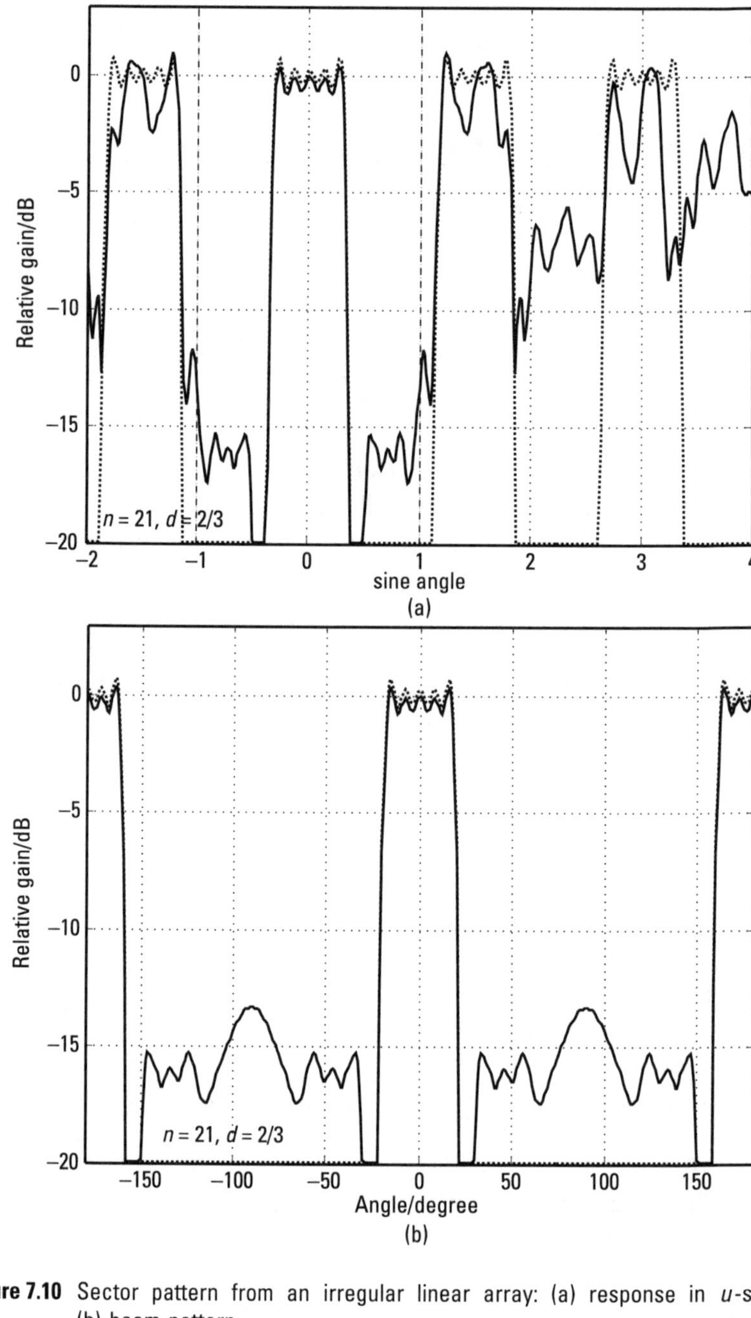

Figure 7.10 Sector pattern from an irregular linear array: (a) response in *u*-space; (b) beam pattern.

separation d wavelengths, by a pseudorandom step chosen within an interval of width $d - 0.5$, which ensures that the elements are at least half a wavelength apart. Figure 7.10(a) shows the response in u space for an array of 21 elements at an average spacing of 2/3. A sector beam of width 40 degrees centered at broadside was specified. A regular array would have a pattern repetitive at an interval of 1.5 in u, and this is shown by the dotted response. The irregular array "repetitions" are seen to degrade rapidly, but the pattern that matters is that lying in the interval [−1, 1] in u. This part of the response leads to the actual pattern in real space, shown in Figure 7.10(b). We note that the side lobes are up to about −13 dB, rather poorer than for the patterns from regular arrays shown in Figures 7.6, 7.8, and 7.9, though this level varies considerably with the actual set of element positions chosen. The integration interval I was chosen to be [−1, 1], to give the least squared error solution over the full angle range (from −90 degrees to +90 degrees, and its reflection about the line of the array).

A second example is given in Figure 7.11 for an array of 51 elements, but illustrating the effect of steering. In Figure 7.11(a, b) the 40-degree beam is steered to 10 degrees, and again we see the rapid deterioration of the approximate repetitions in u space of the beam, and a nonsymmetric side-lobe pattern, though the levels are roughly comparable with those of the first array. The average separation is 0.625 wavelengths, giving a repetition interval of 1.6 in u. If we steer the beam to 30 degrees [Figure 7.10(c, d)], there is a marked deterioration in the beam quality. This is because one of the repetitions falls within the interval I over which the pattern error is minimized, so the part of this beam (near $u = -1$) that should be zero is reduced. At the same time the corresponding part of the wanted beam (near $u = \frac{1}{2}$) should be unity, so the solution tries to hold this level up. We note that the levels end up close to −6 dB, which corresponds to an amplitude of 0.5, showing that the error has been equalized between these two requirements. We note from the dotted responses that the result would be much the same using a regular array. In fact, this problem would be avoided by choosing I to be of width 1.6 (the repetition interval) instead of 2, preserving the quality of the sector beam, but in this case the large lobe around −90 degrees would be the full height, near 0 dB. Even if this solution (with a large grating lobe) were acceptable for the regular array, it is not so satisfactory for the irregular array as the distorted repetitions start to spread into the basic least squares estimation interval, as the array becomes more irregular, creating more large side lobes.

Thus, although a solution can be found for the irregular array, its usefulness is limited for two reasons; the set of nonorthogonal exponential

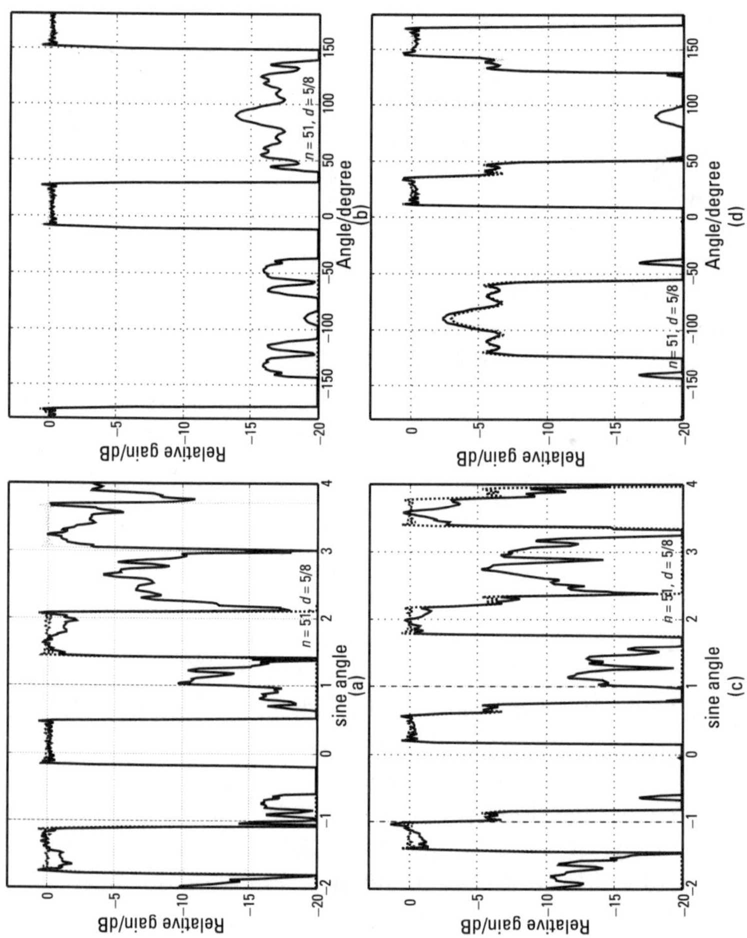

Figure 7.11 Sector patterns from a steered irregular linear array: (a) response in u-space, beam at 10 degrees; (b) beam pattern, beam at 10°; (c) response in u-space, beam at 30°; (d) beam pattern, beam at 30°.

functions (from the irregular array positions) used to form the required pattern is not as good as the set used in the regular case, and if the element separation is to be 0.5 wavelength as a minimum, an irregular array must have a mean separation of more than 0.5 wavelength, leading to grating (or approximate grating) effects.

7.5 Summary

As there is a Fourier transform relationship between the current excitation across a linear aperture and the resultant beam pattern (in terms of u, a direction cosine coordinate), there is the opportunity to apply the rules-and-pairs methods for suitable problems in beam pattern design. This has the now familiar advantage of providing clarity in the relationship between aperture distribution and beam patterns, where both are expressed in terms of combinations of relatively simple functions.

However, there is the complication to be taken into account that the "angle" coordinate in this case is not the physical angle but the direction cosine along the line of the aperture. In the text we have taken the angle θ to be measured from broadside to the aperture, and defined the corresponding Fourier transform variable u as $\sin\theta$, so that $u = \cos(\pi/2 - \theta)$, the cosine of the angle measured from the line along the aperture. In this u domain, beam shapes remain constant as beams are steered, while in real space they become stretched out when steered towards the axis of the aperture. Furthermore, the transform of the aperture distribution produces a function that can be evaluated for all real values of u, but only the values of u lying in the range -1 to 1 correspond to real directions.

Both continuous apertures and discrete apertures can be analyzed, the latter corresponding to ideal antenna arrays, with point, omnidirectional, elements. In this chapter we have concentrated on the discrete, or array, case. The regular linear array, which is very commonly encountered, is particularly amenable to the rules-and-pairs form of analysis. In this case, the regular distribution (a comb function) produces a periodic pattern in u space (a rep function). In the case of a directional beam, the repetitions of this beam are potential grating lobes, which are generally undesirable, but if the repetition interval is adequate, there will be no repetitions within the basic interval in u corresponding to real space and hence no grating lobes. The condition for this (that the elements be no more than half a wavelength apart) is very easily found by this approach. Two variations on the directional beam for producing different low side-lobe patterns were studied in Section

7.3.1. These exercises, whether or not leading to useful solutions for practical application, are intended to illustrate how the rules-and-pairs methods can be applied to achieve solutions to relatively challenging problems with quite modest effort. It was seen in Section 7.3.3 that very good beams covering a sector at constant gain can be produced, again very easily, using the rules-and-pairs method.

The case of irregular linear arrays can also be tackled by these methods. However, the rules-and-pairs technique is not appropriate for finding directly the discrete aperture distribution that will give a specified pattern when the elements are irregularly placed. Instead, the problem is formulated as a least squared error match between the pattern generated by the array and the required one. In this case, the discrete aperture distribution is found to be the solution of a set of linear equations, conveniently expressed in vector-matrix form. The elements of both the vector and the matrix are obtained as Fourier transform functions evaluated at points defined by the array element positions. Again the sector pattern problem was taken and it was shown that this approach gives the same solution as that given directly by the Fourier transform in the case of the regular array, confirming that this solution is indeed the least squared error solution. For the irregular array, we obtain sector patterns as required, though with perhaps higher side-lobe levels and with some limitations on the array (not too irregular or too wide an aperture) and on the angle to which the beam can be steered away from broadside. These limitations are not weaknesses of the method, but a consequence of the irregular array structure that makes achieving a given result more difficult.

Final Remarks

The illustrations of the use of the rules and pairs technique in Chapters 3 to 7 show a wide range of applications and how some quite complex problems can be tackled using a surprisingly small set of Fourier transform pairs. The method seems to be very successful, but on closer inspection we note that the functions handled are primarily amplitude functions—the only phase function is the linear phase function due to delay. Topics such as the spectra of chirp (linear frequency modulated) pulses or nonlinear phase equalization have not been treated, as the method, at least as at present formulated, does not handle these. There may be an opportunity here to develop a similar calculus for these cases.

A considerable amount of work, in Chapters 5 and 6, is directed at showing the benefits of oversampling (only by a relatively small factor in some cases) in reducing the amount of computation needed in the signal processing under consideration. As computing speed is increasing all the time, it is sometimes felt that little effort should go into reducing computational requirements. However, apart from the satisfaction of achieving a more elegant solution to a problem, there may be good practical reasons. Rather analogously to C. Northcote Parkinson's law, "work expands so as to fill the time available for its completion," there seems to be a technological equivalent: "user demands rise to meet (or exceed) the capabilities of equipment." While at any time an advance in speed of computation may enable current problems to be handled comfortably, allowing the use of inefficient implementations, requirements will soon rise to take advantage of the increased performance—for example, higher bandwidth systems, more real-time processing, and more comprehensive simulations. Cost could also be a significant factor, particular for real-time signal processing—it may well be much more economical to put some theoretical effort into finding an efficient implementation on lower performance equipment than require expensive equipment for a more direct solution, or alternatively to enable the processing to be carried out with less hardware.

Finally, while it is tempting to use simulations to investigate the performance of systems, there will always be a need for theoretical analysis to give a sound basis to the procedures used and to clarify the dependence of the system performance on various parameters. In particular, analysis will define the limits of performance, and if practical equipment is achieving results close to the limit, it is clear that little improvement is possible and need not be sought; on the other hand, if the results are well short of the limit, then it is clear that substantial improvements may be possible. The Fourier transform (now incorporating Fourier series) is a valuable tool for such analysis, and as far as Woodward's rules and pairs method makes this operation easier and its results more transparent, it is a welcome form of this tool.

About the Author

After earning a degree in physics at Oxford University (where, coincidentally, he was a member of the same college as P. M. Woodward, whose work has been the starting point for this book), David Brandwood joined, in 1959, the Plessey Company's electronics research establishment at Roke Manor—now Roke Manor Research, a Siemens company. Apart from one short break, he has remained there since, studying a variety of electronic systems and earning a degree in mathematics at the Open University to assist this work. His principal fields of interest have been adaptive interference cancellation, particularly for radar; adaptive arrays; superresolution parameter estimation; and, recently, blind signal separation.

Index

Recent Titles in the Artech House Radar Library

David K. Barton, Series Editor

Advanced Techniques for Digital Receivers, Phillip E. Pace

Airborne Pulsed Doppler Radar, Second Edition, Guy V. Morris and Linda Harkness, editors

Bayesian Multiple Target Tracking, Lawrence D. Stone, Carl A. Barlow, and Thomas L. Corwin

Computer Simulation of Aerial Target Radar Scattering, Recognition, Detection, and Tracking, Yakov D. Shirman, editor

Design and Analysis of Modern Tracking Systems, Samuel Blackman and Robert Popoli

Digital Techniques for Wideband Receivers, Second Edition, James Tsui

Electronic Intelligence: The Analysis of Radar Signals, Second Edition, Richard G. Wiley

Electronic Warfare in the Information Age, D. Curtis Schleher

EW 101: A First Course in Electronic Warfare, David Adamy

Fourier Transforms in Radar and Signal Processing, David Brandwood

Fundamentals of Electronic Warfare, Sergei A. Vakin, Lev N. Shustov, and Robert H. Dunwell

Handbook of Computer Simulation in Radio Engineering, Communications, and Radar, Sergey A. Leonov and Alexander I. Leonov

High-Resolution Radar, Second Edition, Donald R. Wehner

Introduction to Electronic Defense Systems, Second Edition, Filippo Neri

Introduction to Electronic Warfare, D. Curtis Schleher

Introduction to Electronic Warfare Modeling and Simulation, David L. Adamy

Microwave Radar: Imaging and Advanced Concepts, Roger J. Sullivan

Modern Radar System Analysis, David K. Barton

Multitarget-Multisensor Tracking: Applications and Advances Volume III, Yaakov Bar-Shalom and William Dale Blair, editors

Principles of High-Resolution Radar, August W. Rihaczek

Principles of Radar and Sonar Signal Processing, François Le Chevalier

Radar Cross Section, Second Edition, Eugene F. Knott et al.

Radar Evaluation Handbook, David K. Barton et al.

Radar Meteorology, Henri Sauvageot

Radar Reflectivity of Land and Sea, Third Edition, Maurice W. Long

Radar Resolution and Complex-Image Analysis, August W. Rihaczek and Stephen J. Hershkowitz

Radar Signal Processing and Adaptive Systems, Ramon Nitzberg

Radar System Performance Modeling, G. Richard Curry

Radar Technology Encyclopedia, David K. Barton and Sergey A. Leonov, editors

Range-Doppler Radar Imaging and Motion Compensation, Jae Sok Son et al.

Theory and Practice of Radar Target Identification, August W. Rihaczek and Stephen J. Hershkowitz

Time-Frequency Transforms for Radar Imaging and Signal Analysis, Victor C. Chen and Hao Ling

For further information on these and other Artech House titles, including previously considered out-of-print books now available through our In-Print-Forever® (IPF®) program, contact:

Artech House
685 Canton Street
Norwood, MA 02062
Phone: 781-769-9750
Fax: 781-769-6334
e-mail: artech@artechhouse.com

Artech House
46 Gillingham Street
London SW1V 1AH UK
Phone: +44 (0)20 7596-8750
Fax: +44 (0)20 7630-0166
e-mail: artech-uk@artechhouse.com

Find us on the World Wide Web at:
www.artechhouse.com